INTEGRATED PEST MANAGEMENT FOR
CITRUS

UNIVERSITY OF CALIFORNIA
STATEWIDE INTEGRATED PEST MANAGEMENT PROJECT
DIVISION OF AGRICULTURE AND NATURAL RESOURCES

PUBLICATION 3303
1984

PRECAUTIONS FOR USING PESTICIDES

Pesticides are poisonous and must be used with caution. READ THE LABEL BEFORE OPENING A PESTICIDE CONTAINER. Follow all label precautions and directions, including requirements for protective equipment. Use a pesticide only against pests specified on the label or in published University of California recommendations. Apply pesticides at the rates specified on the label or at lower rates if suggested in this publication. Laws, regulations, and information concerning pesticides change frequently, so be sure the publication you are using is up to date.

Legal Responsibility. The user is legally responsible for any damage due to misuse of pesticides. Responsibility extends to effects caused by drift, runoff, or residues.

Transportation. Do not ship or carry pesticides together with food or feed in a way that allows contamination of the edible items. Never transport pesticides in a closed passenger vehicle or in a closed cab.

Storage. Keep pesticides in original containers until used. Store them in a locked cabinet, building, or fenced area where they are not accessible to children, unauthorized persons, pets, or livestock. DO NOT store pesticides with foods, feed, fertilizers, or other materials that may become contaminated by the pesticides.

Container Disposal. Dispose of empty containers carefully. Never reuse them. Make sure empty containers are not accessible to children or animals. Never dispose of containers where they may contaminate water supplies or natural waterways. Consult your county agricultural commissioner for correct procedures for handling and disposal of large quantities of empty containers.

Protection of Non-Pest Animals and Plants. Many pesticides are toxic to useful or desirable animals, including honeybees, natural enemies, fish, domestic animals, and birds. Crops and other plants may also be damaged by misapplied pesticides. Take precautions to protect non-pest species from direct exposure to pesticides and from contamination due to drift, runoff, or residues. Certain rodenticides may pose a special hazard to animals that eat poisoned rodents.

Posting Treated Fields. For some materials, re-entry intervals are established to protect field workers. Keep workers out of the field for the required time after application and, when required by regulations, post the treated areas with signs indicating the safe re-entry date.

Harvest Intervals. Some materials or rates cannot be used in certain crops within a specified time before harvest. Follow pesticide label instructions and allow the required time between application and harvest.

Permit Requirements. Many pesticides require a permit from the county agricultural commissioner before possession or use. When such materials are recommended in this publication, they are marked with an asterisk (*).

Processed Crops: Some processors will not accept a crop treated with certain chemicals. If your crop is going to a processor, be sure to check with the processor before applying a pesticide.

Crop Injury. Certain chemicals may cause injury to crops (phytotoxicity) under certain conditions. Always consult the label for limitations. Before applying any pesticide, take into account the stage of plant development, the soil type and condition, the temperature, moisture, and wind. Injury may also result from the use of incompatible materials.

Personal Safety. Follow label directions carefully. Avoid splashing, spilling, leaks, spray drift, and contamination of clothing. NEVER eat, smoke, drink, chew while using pesticides. Provide for emergency medical care IN ADVANCE as required by regulation.

To simplify information, trade names of products have been used. No endorsement of named products is intended, nor is criticism implied of similar products which are not mentioned.

ORDERING PROCEDURE

Write to: Division of Agriculture and
Natural Resources Publications
University of California
Berkeley, California 94720

Include publication number 3303.

Within the United States: Make check or money order payable to the Regents of the University of California. Price includes postage and handling charges. California residents must include sales tax or furnish a resale number. Orders must be accompanied by payment or a formal purchase order.

Outside the United States: Price does NOT include postage, handling. Please do not send payment; request a Pro Forma Invoice, which will include postage and handling. Indicate the number of copies desired and whether they should be shipped by surface or air mail. Please be prepared to make your payment in U.S. dollars, through a bank located in the United States.

Note: A 20 percent discount is given on orders of 10 or more copies sent to one address. All sales are final; no publication may be returned for cash or credit.

No part of this publication may be reproduced, stored in a retrieval system, or transmitted, in any form or by any means, electronic, mechanical, photocopying, recording, or otherwise, without the written permission of the publisher and the author.

© 1984 by The Regents of the University of California

ISBN: 0-931876-65-6
Library of Congress No. 83-082076

Printed in the United States of America

The University of California in compliance with the Civil Rights Act of 1964, Title IX of the Education Amendments of 1972, and the Rehabilitation Act of 1973 does not discriminate on the basis of race, creed, religion, color, national origin, sex, or mental or physical handicap in any of its programs or activities, or with respect to any of its employment policies, practices, or procedures. The University of California does not discriminate on the basis of age, ancestry, sexual orientation, marital status, citizenship, nor because individuals are disabled or Vietnam-era veterans. Inquiries regarding this policy may be directed to the Affirmative Action Officer, 2120 University Ave., University of California, Berkeley, California 94720 (415) 644-4270.

4½m-2/84 PRINTED FEBRUARY 1984

Contributors and Acknowledgments

This manual was produced under the auspices of the Statewide IPM Project, James M. Lyons, Director, Howard Ferris, Associate Director.

Prepared by the IPM Manual Group of the Statewide IPM Project, U.C. Davis, Mary Louise Flint, Director.

Brunhilde Kobbe, Senior Writer
Jack Kelly Clark, Principal Photographer

Technical Coordinators

John E. Pehrson, Subtropical Horticulture Specialist, Cooperative Extension, Lindcove Field Station

Donald L. Flaherty, Entomology Farm Advisor, Cooperative Extension, Tulare County

Neil V. O'Connell, Farm Advisor, Cooperative Extension, Tulare County

Phil A. Phillips, Area IPM Specialist, Cooperative Extension, Ventura County

Contributors

Entomology: E. Laurence Atkins, Jr., J. Blair Bailey, O.L. Brawner, Robert D. Brown, Paul DeBach, T.W. Fisher, D.L. Flaherty, Charles Kennett, Robert F. Luck, James A. McMurtry, Daniel S. Moreno, J.G. Morse, N.V. O'Connell, P.A. Phillips, Louis A. Riehl, Mike Rose, Lynell Tanigoshi

Horticulture: Tom W. Embleton, David Goldhamer, Carol J. Lovatt, C. Dean McCarty, John E. Pehrson, Robert G. Platt

Nematology: Sahag Garabedian, Seymour van Gundy, Reinhold Mankau, John D. Radewald

Pesticide Application: Louis A. Riehl, John E. Pehrson, Glenn Carman

Plant Pathology: David J. Gumpf, John A. Menge, Howard D. Ohr, John E. Pehrson

Vertebrates: Rex E. Marsh, William Clark

Weed Science: Bill B. Fischer, Lowell S. Jordan

Special thanks to:

R. Allert, D. Anderson, D. Avis, J. Baritelle, H. Bender, W.P. Bitters, R.W. Brazelton, R.M. Burns, W. Carmean, C.W. Coggins, Jr., G.L. Dickey, E.J. Dietrick, J.A. Dodds, H.S. Elmer, W.H. Ewart, C. Farrar, J. Gorden, L. Graebner, H.J. Griffith, P. Haney, W.H. Krueger, S. Lindow, B. Lingren, G. Loughner, J. Maranto, J. Moffitt, C. Musgrove, E. Mussen, J.L. Pappas, B. Peterson, P.A. Rude, M.W. Stimmann, J.R. Stewart, S. Warner, C. Wood, M.L. Zavala for suggestions, manuscript review, or help with photographs.

Information in this manual has been derived from research supported by the University of California, the USDA, and the Citrus Research Board. Thanks to Agricultural Sciences Publications, University of California, for making available editors and artists.

Special appreciation goes to J. Blair Bailey, Jim Gorden, and the Technical Coordinators for assistance in coordinating numerous photo sessions.

Production

Manuscript Preparation: Betty Rudd
Editing: Margaret Klein
Drawings: Pamela Fabry, Marvin Ehrlich
Design and Production Coordination: Naomi Schiff

Table of Contents

Integrated Pest Management for Citrus ... 6

The Citrus Tree: Development
and Growth Requirements ... 8
    Seasonal Cycle of Citrus ... 8
    Growth Requirements ... 10

Managing Pests in Citrus ... 12
  Pest Identification ... 12
  Field Monitoring ... 12
    Monitoring Pests ... 13
    Monitoring Weather ... 13
    Accumulating Degree-Days ... 14
  Control Action Guidelines ... 16
  Management Methods ... 16
    Scion Cultivar and Rootstock Selection ... 16
    Soil and Water Management ... 17
        Irrigation Methods–Scheduling Irrigations
    Fertilizing ... 22
    Frost Protection ... 23
    Harvest ... 24
    Pruning ... 25
    Ground Cover ... 25
    Biological Control ... 25
    Pesticides ... 27
        Pesticide Application–Problems
        Associated With Pesticide Use

Vertebrates ... 39
    Management Methods ... 39
  Pocket Gophers ... 41
  California Ground Squirrel ... 42
  Meadow Mice ... 45
  Black-Tailed Jackrabbits ... 46
  Other Vertebrates in Citrus ... 47

Insects, Mites and Snails ... 49
    The San Joaquin Valley ... 49
    The Coastal-Intermediate and
    Interior Districts ... 50
    The Desert Valleys ... 50

  Monitoring Insects and Mites ... 50
    Monitoring Methods ... 55
  Scale Insects ... 57
    California Red Scale ... 57
    Purple Scale ... 65
    Citricola Scale ... 66
    Brown Soft Scale ... 68
    Black Scale ... 69
    Cottonycushion Scale ... 70

Citrus Thrips ... 72
Orangeworms ... 77
    Monitoring Orangeworms ... 77
  Key to Orangeworms in California Citrus ... 78
  Citrus Cutworm ... 80
  Fruittree Leafroller ... 82
  Orange Tortrix ... 84
  Omnivorous Leafroller ... 85
  Western Tussock Moth ... 86
  Amorbia ... 87
  Pink Scavenger Caterpillar ... 89
  Black Anise Swallowtail ... 89
  Citrus Looper ... 89
Mites ... 90
  Citrus Red Mite ... 90
  Twospotted Mite ... 94
  Broad Mite ... 95
  Citrus Bud Mite ... 96
  Citrus Rust Mite ... 96
  Other Mites in Citrus ... 98
    Flat Mite ... 98
    Yuma Spider Mite ... 98
    Sixspotted Mite ... 98
    Lewis Spider Mite ... 98
Ants ... 99
Katydids ... 100
Mealybugs ... 101
Whitefly ... 102
Other Insect Pests ... 105
    Potato Leafhopper ... 105
    Aphids ... 105

Fuller Rose Beetle	106
Grasshoppers	106
Fruit Flies	107
Brown Garden Snail	108

Diseases 110
    Monitoring and Diagnosis of
    Citrus Diseases 111
    Prevention and Management 111

Symptoms on Roots 111
    Phytophthora Root Rot 112
    Dry Root Rot 112
    Armillaria Root Rot 113

Symptoms on the Trunk 114
    Phytophthora Gummosis 114
    Exocortis 115
    Psorosis 115
    Other Trunk Diseases and Disorders 115
        Shell Bark and Dry Bark 115
        Bud Union Disorders 117
        Sunburn 117

Symptoms on Fruit 117
    Brown Rot 117
    Alternaria Rot 118
    Septoria Spot 118
    Other Diseases and Disorders of Fruit 118
        Anthracnose Tearstain 119
        Blue and Green Mold Disease 119
        Botrytis Rot 119
        Chimeras 119
        Frost Damage 119
        Rind Stipple of Grapefruit 119
        Sunburn 119
        Wind Injury 119
        Spray Injury 119

Symptoms on Leaves and Twigs
    Citrus Blast 121
    Botrytis Rot 122
    Chimeras 122
    Twig Dieback 122
    Wind Injury 122
    Mesophyll Collapse 122
    Frost Damage 122
    Spray Injury 122
    Mineral Deficiencies and Toxicities 123

Symptoms Affecting Growth Habit and Yield 125
    Stubborn Disease 125
    Tristeza Disease Complex 126
    Lemon Sieve Tube Necrosis 128

Nematodes 129
    Description and Biology 129
    Damage 130
    Guidelines for Managing Nematodes 130

Weeds 132
    Guidelines for Managing Weeds 133
        Prevention 133
        Control Methods 133
        Monitoring and Control Program 135
    Weed Species Common in Citrus Orchards 137
    Bermudagrass 137
    Dallisgrass 138
    Field Bindweed 138
    Nutsedges 139
    Johnsongrass 140
    Bearded Sprangletop 141
    Barnyardgrass 141
    Spotted Spurge 142
    Turkey Mullein 142

References 143

Glossary 144

Integrated Pest Management for Citrus

The purpose of this manual is to help growers and pest control advisors apply the principles of integrated pest management, or IPM, to California citrus crops. IPM emphasizes preventive methods that provide economical, long-term solutions to pest problems. Pesticides are used only when they are necessary to prevent imminent crop loss or damage; thus IPM strategies minimize hazards to human health and the environment.

Citrus is grown in four major areas of the state: the San Joaquin Valley, the coastal-intermediate district, the interior district, and the desert valleys of southern California; an additional small growing area lies in the northern Sacramento Valley (Figure 1). The regions differ in climate, cultivars grown, and pest problems. In the San Joaquin Valley, where more than half of the 277,000 acres of California citrus are grown, summers are hot and dry, and winters are generally cold and wet. The coastal-intermediate district has a milder climate due to the influence of marine air. The interior district, which is cut off from the coast by mountain ridges, tends to be warmer and dryer in the summer and colder in the winter than the coast. In the desert valleys—the Coachella Valley and the Imperial Valley—temperatures fluctuate widely between day and night, and the humidity is low most of the year.

Climate determines to a large degree the choice of cultivar, the pest problems, and the management options available. Among the major cultivars, navel orange is predominant in the San Joaquin Valley, Valencia orange in coastal-intermediate and interior areas, lemon on the coast, and grapefruit in the desert. All areas, however, have some acreage of nearly all major cultivars. Biological control of insect and mite pests is generally more effective in coastal-intermediate areas than in the San Joaquin Valley and desert. Fungal and bacterial diseases cause more losses in the central valley with its long, wet winters than in other areas.

The nature of the cultivar itself has some influence on cultural practices and pest management actions. Valencia fruit and coastal grapefruit take longer to mature and are exposed longer to pests or environmental stresses than the fruit of other cultivars. Coastal lemons set fruit throughout most of the year; the presence of fruit at various stages of development presents special problems

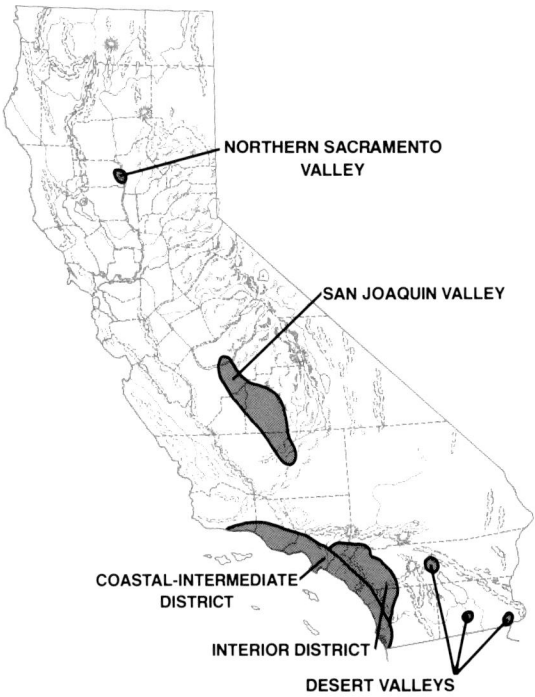

Figure 1. Major citrus-growing areas of California. Almost half of the total citrus acreage is grown in the San Joaquin Valley (139,700 acres), followed by the coastal-intermediate (79,070) and interior (33,590) districts. Smaller growing areas are the desert valleys (Coachella Valley, 18,270; Imperial Valley, 4,500) and the Northern Sacramento Valley (1,800). Acreage figures are from 1981.

to pest management. The development and management of grapefruit and lemon in the desert and other minor cultivars, such as mandarins, tangelos, tangerines, and pummelos, are similar to navel oranges. Limes are similar in development and management to lemons.

Use this book to plan an IPM strategy for your orchard. The introductory chapters on citrus development, growth requirements, general management practices, and monitoring tools provide the background upon which the management guidelines in the pest sections are based. The introductions to the insect, weed, disease, nematode, and vertebrate chapters tell you where and when major pests occur. Detailed descriptions and photographs of pests and damage symptoms are presented later in each chapter. These sections also discuss how you can enhance natural control factors, design a monitoring program, and use control actions most effectively.

IPM is a flexible, evolving strategy that will be updated periodically as new information becomes available. Check regularly with your farm advisor for new developments. Because pesticide registrations change frequently, no specific recommendations are given here. Use the manual together with the latest University of California *Treatment Guide for California Citrus Crops*.

The Citrus Tree: Development and Growth Requirements

The objective of citrus production is to harvest a large crop of high quality fruit. A high quality fruit has a rich flavor resulting from a balance between sweetness and tartness, has firm, juicy flesh, an aroma and size characteristic of the cultivar and, for fresh market fruit, a smooth, deep colored and blemish-free rind. Optimal production can be achieved only with healthy trees. Maintaining tree health requires a basic understanding of the seasonal cycle, the crop's specific requirements, and the impact that pest organisms and cultural or environmental factors have on tree and fruit growth. Without this understanding, it is easy to overlook stress symptoms caused by environmental factors or to confuse them with pest damage. An even more important aspect of IPM is the evaluation of cultural practices and pest control methods for their impact on the total orchard system as well as for their effectiveness in eliminating the primary pest problem.

The Seasonal Cycle of Citrus

In the subtropical climate of California, trees of all citrus cultivars except coastal lemons stop growing during the winter. During this period, the tree maintains a base level of water transport and starch consumption. The main growth flush appears in late February and March. Some additional growth flushes are produced in the summer and fall. Leaves stay on the tree for 1 to 2 years. They are replaced continually, although leaf drop is greatest during the spring flowering period. Environmental factors, such as high temperature, wind, low soil moisture, low relative humidity, nutrient deficiencies, high soil salinity, and pest problems, can trigger premature leaf drop. The photographs on the back cover illustrate the seasonal development of a major cultivar, navel orange.

Most citrus cultivars produce flowers in the spring. Coastal lemons bloom throughout the year but produce flowers more heavily in the spring. With other cultivars, disease, rainfall, or a dry spell followed by irrigation may trigger irregular or "off" bloom. Fruit developing from "off" bloom are usually of inferior quality. Many cultivars, particularly the Valencia and certain mandarins, have an

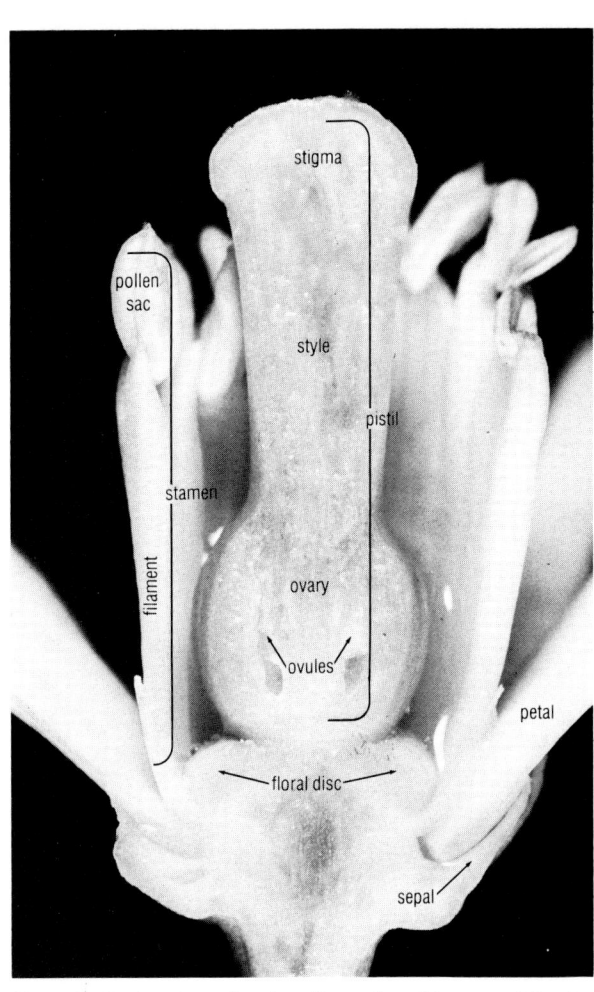

Figure 2. Section through a citrus flower. In cultivars requiring fertilization, sperm cells from the pollen fertilize the egg cells in the ovules, which then develop into seeds. After bloom, the petals, stamen, and style are shed and the ovary develops into the fruit.

alternate bearing habit, setting a heavy crop one year and a light crop the following year.

Flowers are borne on flowering branches called inflorescences. Inflorescences develop in leaf axils on shoots of preceding growth flushes. Inflorescences may bear one to many flowers and be leafless, or they may have one to many leaves. A flower has male structures (stamens) and a single female structure (pistil) (Figure 2). Most cultivars produce viable pollen, although some important commercial cultivars, such as navel orange and Satsuma mandarin, do not. During pollination, pollen grains fall onto the stigma and germinate, producing pollen tubes that grow through the style into the ovules in the ovary. Each pollen tube contains a sperm cell, which fertilizes an egg cell in the ovule, resulting in a zygotic embryo. After fertilization, the ovules develop into seeds.

In most citrus cultivars, one or more asexual embryos will arise from nucellar tissue of a fertilized ovule. The nucellar embryos often crowd out the sexual embryo. Nucellar embryos are valuable in producing clones, which are free of viral diseases the mother plant may carry. Such nucellar clones are used in breeding, providing uniform rootstock material or restoring vigor to bud lines. Several seedlings may grow from a polyembryonic seed, but usually one seedling dominates and survives.

Most citrus cultivars are self-compatible, that is, they can be fertilized by their own pollen; Clementine mandarin, however, is self-incompatible and requires pollination by another mandarin or tangelo cultivar to set fruit. Most other citrus cultivars can produce fruit without fertilization and seed formation, but pummelo and some mandarins and tangelo hybrids set a better crop when pollinated by certain other cultivars. The number of seeds produced depends on the cultivar; navel orange and Satsuma mandarin have little or no viable pollen and few fertile eggs, so no seeds generally form.

Citrus usually blooms abundantly but most flowers and young fruit drop ("early drop"). A combination of environmental and physiological factors seems to determine which flowers develop into fruit that persist to harvest. Persistent flowers are generally those borne on inflorescences with a low flower-to-leaf ratio, that is, on "leafy" inflorescences. Inflorescences whose flowers open late in the bloom period tend to have a lower flower-to-leaf ratio than inflorescences with early opening flowers, and more late-maturing flowers produce fruit that survive to harvest. In addition, faster-growing ovaries have a better chance to set fruit and survive to harvest, whereas slower-growing fruit are more likely to drop off the tree at an early stage of development. With more detailed knowledge of the parameters governing fruit set and persistence, it may be possible to predict crop yield much earlier and allow a better cost/benefit analysis of early season pest management actions, such as control of orangeworms and citrus thrips.

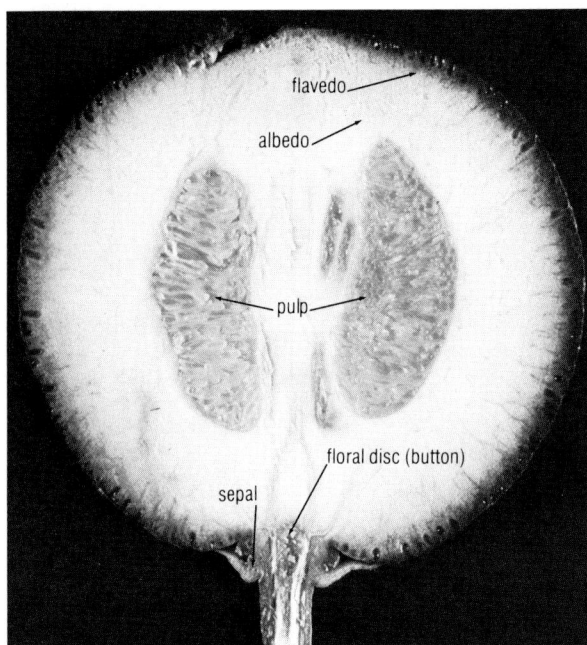

Figure 3. Section through a young navel orange. The rind consists of a thin, outer layer (flavedo), which contains pigments and oil cells, and a thick, white, spongy inner layer (albedo). The pulp, composed of juice vesicles, is just beginning to enlarge.

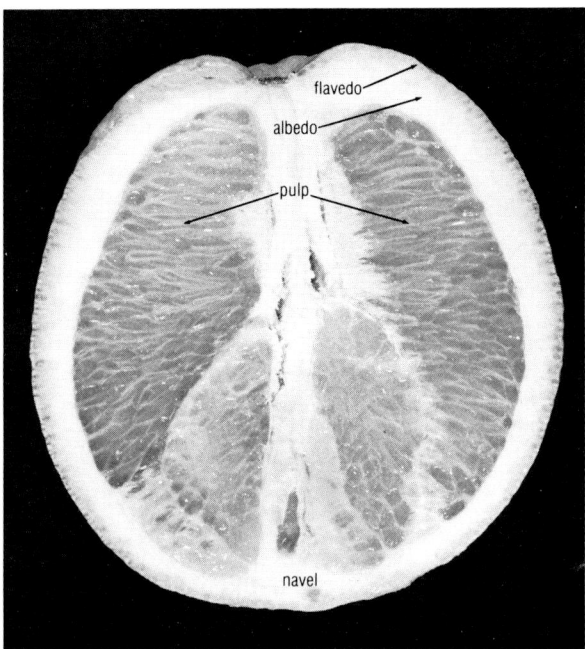

Figure 4. Section through a mature navel orange. The juice vesicles are now greatly enlarged and filled with juice of a certain acid/sugar ratio; the rind has become thinner as the fruit matured. The green pigments of the flavedo will have changed to bright orange if night temperatures have been low.

After bloom, the young fruit undergoes rapid cell division for up to 9 weeks, depending on the temperature. High temperatures occurring between mid-May and mid-July cause fruit to stop growing and abscise ("June drop"). Growth during this time is due to increasing peel thickness. After this stage, the cells enter a period of rapid cell enlargement. The peel, including the outer, colored flavedo and inner, white albedo layers, reaches maximum thickness early in this period, becoming thinner as the pulp continues to increase in size (Figures 3, 4). A maturation period follows. The fruit continue to grow for as long as they are left on the tree, but they grow at a very reduced rate. During the maturation stage, the juice with its acids, mainly citric acid, starts to form. In oranges, grapefruit, and mandarins, sugars begin to accumulate while the acid content is decreasing.

Unlike deciduous fruit, citrus fruit have no clearly identifiable point of maturity. Color can serve as an approximate guide but is generally not reliable because color development depends largely on temperature, especially on low night temperature, and the mineral nutrition of the tree. For oranges, mandarins, and grapefruit, the ratio of soluble solids—mainly sugars—to acids is more important for determining legal maturity than color. The fruit continue to improve in flavor and taste for several weeks or even months. Thus, citrus fruit store well on the tree. Certain hormone sprays can even prolong tree storage (see Harvest, p. 24). Later, taste declines due to a breakdown of acid and flavor components. Lemons are considered ripe when they have reached a certain percentage of juice per volume fruit.

Growth Requirements

Tree growth and fruit development require adequate light, water, and nutrients, and appropriate temperatures. Light is essential for photosynthesis, the process by which green plants manufacture sugars, their major food source (Figure 5). The plant may metabolize the sugars immediately, store them as starch for later use, or concentrate them in certain plant parts, such as the fruit. Sugars also serve as building blocks for thousands of other compounds, including pigments, flavor components, and structural molecules like cellulose.

All basic plant functions require water. Only a small fraction of the water taken up by the tree is retained; most of it passes through the plant and out leaf pores (stomata) during the process called transpiration. Transpiration is necessary to supply water to the leaves for photosynthesis, to carry nutrients to all parts of the tree, and to cool the plant. Climatic factors, such as the amount of sunshine, humidity, heat, and wind, affect the rate of transpiration.

Various nutrients are needed for good tree growth and fruit production. The nutrients needed in largest amounts, often called macronutrients, are nitrogen, phosphorus, potassium, calcium, magnesium, and sulfur. Micronutrients are needed only in small amounts but are essential to basic tree function and crop production. These include zinc, manganese, iron, boron, molybdenum, copper, and chlorine. California orchard soils contain sufficient amounts of most essential nutrients either because they are already present in the soil or because they are returned periodically through the decay of organic matter or as impurities in irrigation water. Others, mainly nitrogen but also zinc and manganese, have to be replenished through fertilization.

Beneficial fungi, called mycorrhizae, greatly enhance the uptake of nutrients and water. Mycorrhizal fungi live in a symbiotic relationship with many plant roots, including citrus roots, providing them with water and many essential nutrients, such as phosphorus, zinc, and copper, while receiving carbohydrates from the plant roots. They are efficient suppliers because their threadlike hyphae explore much larger areas than the root hairs. Mycorrhizae are sensitive to preplant soil fumigants used against fungal soil pathogens. A scheme for mass-producing the beneficial fungi and inoculating citrus seedlings is used experimentally in some nurseries. Using mycorrhizae can substantially reduce the amount of fertilizer needed and improve the growth of nursery trees. Research is underway to find ways to protect the mycorrhizae under orchard conditions and enhance their beneficial action.

Citrus produces best under a mild subtropical climate. Wide fluctuations between day and night temperatures promote acid formation, resulting in a rich flavor. Cool night temperatures also trigger the development of the bright orange or yellow color of the rind. Total heat accumulation determines the amount of sugar produced and thus the time it takes for oranges and grapefruit to ripen. Fruit grown in the desert ripens first, followed by those grown in the central valley and the intermediate areas; fruit on the coast ripens last. Citrus cultivars differ in their heat requirements. Grapefruit cultivars require more heat than other cultivars, and in California they develop best flavor in the desert regions. Acidic citrus, such as lemon and lime, do not need to sweeten, so they grow well in the mild climate of the coastal areas. Citrus is sensitive to extreme temperatures, which can cause sunburn or frost damage (see pages 119 and 122).

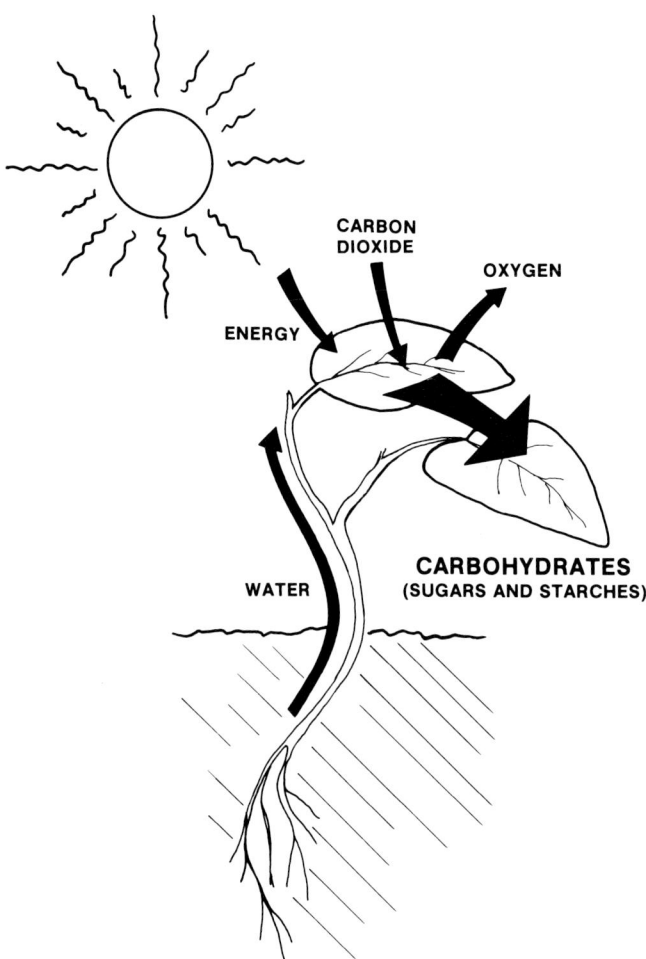

Figure 5. In photosynthesis, green plants use energy from the sun to convert carbon dioxide from the air and water from the soil into sugars, their primary food source. In the process, oxygen is released into the air.

Managing Pests in Citrus

Integrated pest management treats pests as part of a crop production system that includes not only the crop and its pests, but also the physical and biological environment in which the crop is grown. A good IPM program coordinates pest management activities with each other and with cultural operations to achieve economical and long-lasting solutions to pest problems; the emphasis is on anticipating and preventing problems when possible. Figure 6 lists the cultural operations and pest management actions most likely to be needed during the year. This summary sheet also gives the page numbers where you can find details about a particular management practice, and it allows you to check off actions that you have completed during a season.

An integrated pest management program has four major components:

- pest identification;
- field monitoring;
- control action guidelines;
- effective methods for prevention and control.

Pest Identification

Correctly identifying pest species and stress symptoms is fundamental to choosing corrective management strategies. Most pest management tools, including pesticides, are effective only against a select group of species; therefore, different practices may be needed even for closely related species. In some cases, the symptoms caused by pest organisms closely resemble those caused by nutrient deficiencies or other soil problems. The photographs, descriptions, and seasonal charts in this manual are intended to help identify problems. Farm advisors can assist you in confirming identification or identifying unusual problems.

Field Monitoring

Field monitoring provides information on daily or seasonal orchard conditions, such as the status of pests, weather, or soil moisture. This information is used to predict and evaluate potential pest problems. Because conditions vary even between neighboring orchards, individual groves must be monitored. Check regularly the pest species present, the maturity and health of the crop, the weather, plant environment, including soil conditions, and when appropriate, the population levels of pest and beneficial organisms. Keep records of monitoring results, weather, and management activities. During the season, these records show whether pests or natural enemy populations are increasing or decreasing. They also help forecast possible outbreaks or the next generations of a pest. Pest population counts, together with records of control measures, cultural practices and weather conditions, help determine whether future control actions will be needed. Simple tables and graphs of data help define patterns, and maps can identify localized problems and pest distribution. Over the years, these records provide valuable historical data for long-term orchard management.

Monitoring Pests

The frequency of field checking varies with the individual pest species and the season. Monitoring methods for insects include checking a certain number of fruit or leaves, shaking foliage into a sweep net, or using traps. Monitoring diseases may require checking fruit, leaves, or trunks for symptoms or digging up some feeder roots for inspection.

Sampling programs for most species are needed only at certain times of the year. Start sampling well before populations begin to build and continue through the pest's damaging stages. Monitor individual orchards for insect and mite pests and their natural enemies when pests are likely to be present. The sampling frequency varies with the district, but generally, weekly sampling is recommended with more frequent checking when the monitoring program indicates an outbreak.

No weekly sampling programs have been developed to assess populations of weeds, diseases, nematodes, and vertebrates. It is important, however, to note their presence and watch for changes in their status. Prepare a record at least twice a year for each orchard, noting the dominant weeds, diseases, vertebrates and any other significant stresses. Sample for nematodes when you suspect they are causing losses or when the site history indicates they may become a problem. Specific monitoring procedures are described in the individual pest sections.

Monitoring Weather

Weather greatly influences the development of both the citrus tree and its pests. Accurate weather forecasts are necessary for timing pest control activities and cultural

Figure 6. Seasonal check lists of management actions for a citrus orchard

✔	MANAGEMENT ACTION	PAGE
WINTER 19____ (Dec-Feb)		
☐	fertilize (soil applied nitrogen, potassium) if indicated by leaf analysis	22
☐	protect against frost as needed	23
☐	apply growth regulator to navels, if needed	25
☐	harvest navels, mandarins, grapefruit (desert), lemons (San Joaquin Valley)	25
☐	monitor diseases (gummosis, brown rot)	111
☐	survey winter weeds present (use record sheet, Fig. 64)	135
☐	check soil moisture	17
☐	monitor brown garden snail if weather is mild	109
☐	prune navels (every 3-5 years) after harvest and no frost	25
SPRING 19____ (Mar-May)		
☐	set up irrigation schedule, irrigate as needed	17
☐	apply nutritional sprays (zinc, manganese, urea, potassium nitrate)	22
☐	harvest navels, Valencias, grapefruit (desert), lemons (coast)	24
☐	monitor vertebrate activity, control if needed	39
☐	check for root and trunk diseases	111,114
☐	monitor citrus cutworm, fruittree leafroller, control if needed	82,83
☐	monitor citrus red mite (use record sheet, Fig. 61)	93
☐	monitor citrus thrips (use record sheets, Figs. 57, 58), control if needed	74
☐	monitor brown garden snail, control if needed	109
☐	monitor Argentine ant, control if needed	99
☐	monitor katydids, control if needed	101
SUMMER 19____ (Jun-Sep)		
☐	follow irrigation schedule, irrigate as needed	17
☐	harvest lemons, Valencias, and grapefruit (coast)	24
☐	apply growth regulator to coastal Valencias and grapefruit if needed	25
☐	have leaf analysis done	22
☐	prune Valencias (every 3-5 years, after harvest), lemons (yearly)	25
☐	monitor California red scale (use record sheet, Fig. 54), control if needed	63
☐	monitor other scales	65-71
☐	monitor citrus red mite (use record sheet, Fig. 61)	93
☐	monitor other mites	94-98
☐	check for other invertebrate pests (mealybugs, whitefly)	101,102
☐	monitor summer weeds (use record sheet, Fig. 64), control if needed	135
FALL 19____ (Oct-Nov)		
☐	continue irrigation as necessary	17
☐	harvest coastal lemons	24
☐	apply growth regulator on navels, if needed	25
☐	prune grapefruit (3-5 years, after harvest)	25
☐	prepare for frost protection	23
☐	continue monitoring California red scale	63
☐	watch for buildup of citrus red mite, treat if necessary	93
☐	watch for leafhoppers	105
☐	protect against brown rot	118
☐	apply preemergence herbicides	133
☐	check for vertebrate damage, control if needed	39

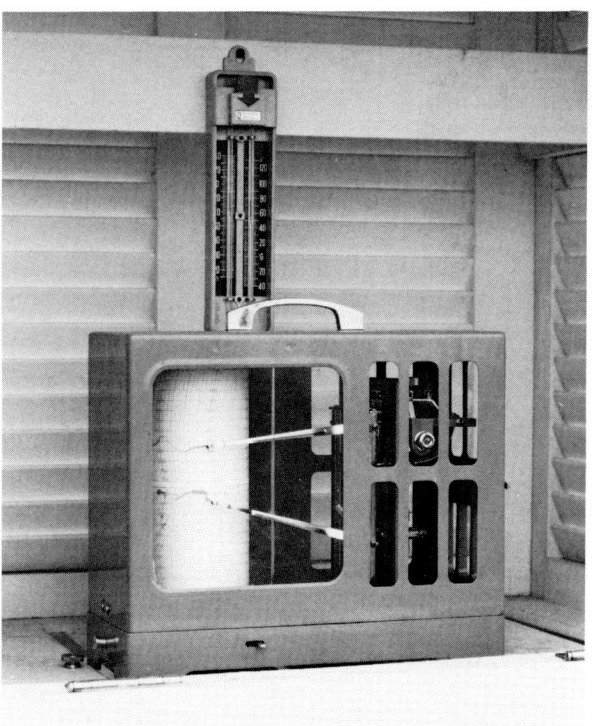

Figure 7. A maximum-minimum thermometer (top) records the highest and lowest temperatures over a 24-hour period. A hygrothermograph (below) records temperature and humidity continuously over several days.

operations, such as irrigation and frost protection. The National Weather Service broadcasts local and regional weather information in California over two very high frequency (VHF) FM bands, 162.40 and 162.55 megahertz. Local newspapers and radio stations report daily high and low temperatures, relative humidity, dew point, and wind conditions, and evapotranspiration data are provided in some areas. The University of California provides a large data base of historical and current weather information to farm advisors through its statewide computer system. Climatological data as well as degree-day calculations are presently available at the terminals in some farm advisor offices, and expansion to more counties is planned.

Weather conditions, particularly temperature and rainfall, may vary considerably within a few miles due to differences in topography, elevation, storm paths, or bodies of water present. To have the most accurate data for your orchard, you may want to set up your own basic weather station. It would include a maximum/minimum thermometer or a thermograph and a precipitation gauge. A maximum/minimum thermometer records the highest and lowest temperature of the day, whereas a thermograph gives a continuous temperature reading for 24 hours or longer (Figure 7). This information is used for calculating degree-days (see below) and assessing frost hazard and winter mortality of important pests, such as the California red scale. A precipitation gauge, which measures the amount of rainfall, is important in setting up an accurate water budget.

For predicting frost hazard accurately, install some thermometers in known cold spots and some in areas with average temperatures. Compare your readings with those of your local Fruit Frost Service station, which usually measures temperatures in unprotected orchards. This will tell you how many degrees above or below the forecast your orchard temperature runs so that you can predict more accurately the frost hazard. Place the thermometers 5 feet high and protect them from direct sunlight. A simple canopy shelter is sufficient for thermometers (Figure 8) but thermographs should be enclosed in a weather shelter (Figure 9). Also have one thermometer outside the orchard and upwind, where it is unaffected by frost protection measures. Maintain the instruments according to the manufacturer's instructions and calibrate them regularly to assure accuracy. Keep records of all your observations.

Accumulating Degree-Days

The concept of degree-days is important for understanding crop and pest development. Research is underway to establish degree-days as a tool for forecasting fruit development and outbreaks of pests such as citrus thrips and California red scale.

The growth rates of a plant and its invertebrate and microbial pests are closely related to temperature. For instance, in a warm year both will develop faster than in a cold year. Therefore, the calendar date alone is not a precise guide for carrying out crop and pest management activities. For more accurate predictions of crop and pest development, time and temperature have to be considered together. The rate of development depends on the amount of heat above a certain developmental threshold the plant or pest experiences during a given time period. Measuring the amount of heat accumulating over time provides a "physiological time scale," which is biologically more meaningful and more useful than calendar days. The unit of the physiological time scale is the degree-day (°D). It is defined as one degree above the developmental threshold maintained for a full day. You can measure the temperature in Fahrenheit or Celsius; all degree-day figures in this manual are based on the Fahrenheit unit.

Each species of plant, invertebrate, or microorganism has a specific lower and upper thermal developmental threshold. The lower threshold is the temperature below which no development occurs. Above this threshold, the rate of development increases with temperature at a roughly linear rate until further growth is inhibited at the upper developmental threshold. An organism's lower and sometimes upper thresholds must both be known to calculate degree-days accurately.

To use degree-days for keeping track of crop or pest development, you must know the number of degree-days (amount of heat units) required to complete each growth stage. At high temperatures, degree-days accumulate faster, and organisms develop faster. For example, the California red scale does not develop below 53°F (11.7°C) and requries 1100°D above this threshold to complete one generation. When 10°D accumulate every day, a generation is completed in 105 days; where 20°D accumulate per day, a generation is completed in only 53 days.

You can estimate degree-days from the daily minimum and maximum temperatures as outlined in Table 1. This formula provides an approximate guide but is less accurate if the minimum temperature is below the lower threshold. More precise formulas are available for programmable hand calculators and computers. The IPM computer of the University of California is programmed to provide degree-day tables for organisms whose developmental threshold temperatures are known.

Table 1. A Simple Way to Calculate Degree-Days

Add daily minimum and maximum temperatures and divide by 2 to get the average daily temperature.	$\dfrac{74°F + 54°F}{2} = 64°F$
Subtract the lower threshold temperature (e.g., 53°F for the California red scale) from the average daily temperature. The result is the number of degree-days accumulated that day.	$64°F - 53°F = 11°D$

Figure 8. A simple shelter can provide protection for your orchard thermometers.

Figure 9. Protect your instruments for measuring temperature and humidity in a vented weather shelter.

Control Action Guidelines

Control action guidelines indicate when to take management actions to avoid crop losses. They are useful only in conjunction with careful field monitoring and accurate pest identification. For several insect and mite pests, numerical thresholds have been developed that indicate economically damaging population levels, and these have been used as a basis for control action guidelines. Theoretically, a threshold level for an insect or mite pest is reached when the economic damage caused by that pest exceeds the cost of control, including the cost of additional applications due to upset of biological control and incidental expenses. For most diseases, weeds, nematodes, vertebrates, and some insects, the guidelines are currently less well defined and may be based on combining information about the extent of pest infestation, the potential for effective and economical control, the history of an orchard or a region, the stage of crop development, or the weather.

Management Methods

The best methods for an IPM program are those that prevent pest outbreaks and provide long-term and economical control. Preventive methods include following quarantine regulations, removing sources of new infection, choosing scion cultivars and rootstocks less susceptible to pest damage, installing an appropriate irrigation system, providing conditions that enhance the activities of natural enemies of pests, and carrying out various management activities correctly. Use pesticides only where preventive measures are not sufficient and careful field monitoring indicates that economic loss is likely. In some cases, mass releases of natural enemies or cultural practices can reduce damaging levels of pests.

Scion Cultivar and Rootstock Selection

The selection of a scion cultivar and a rootstock is possibly the most critical planning decision because it affects management practices, pest problems, and profits throughout the life of the citrus orchard.

The most important scion cultivars grown in California today are the Washington navel and Valencia orange, Eureka and Lisbon lemons, Marsh (seedless) grapefruit, and mandarin types (tangelos, tangerines). The choice depends on climatic conditions and on desired market qualities.

The navel orange is mainly grown in the San Joaquin Valley. The valley's dry, hot summers and cool winters produce an excellent dessert fruit. It has no seeds, firm flesh, and a bright colored rind that separates easily from the flesh. Navel fruit mature in about 9 months. The main harvest is from late fall through early spring, although it may extend into early summer. Ripe fruit store fairly well on the tree, but at times a growth regulator is needed to extend tree storage without loss in fruit quality.

The Valencia orange is grown in the interior and intermediate areas of southern California, although it can produce well in other areas. It yields a high flavor fruit with few seeds, a tough, firm rind, and superior juice qualities. It is valuable for juice products as well as for the fresh market. The Valencia orange matures in 12 to 15 months. Ripe fruit are less likely to drop than navel oranges or grapefruit, and growth regulators are only occasionally used in southern California to extend tree storage. Valencias are harvested from early spring through late fall.

Lemons are grown mainly on the coast, where they bloom most of the year and produce more fruit than in other areas. Mature fruit can be harvested over a 9-month period. Eureka has been the most common cultivar on the coast. It sets fruit mostly at the tips, where they may be exposed to sunburn and wind scarring. The Lisbon lemon is considered to be better adapted to the interior areas and central valley than Eureka. In the interior districts, desert, and central valley, the fruit mature in the fall and winter. The trees are thorny and set many fruit inside the tree.

Grapefruit produce best in the hot, dry climate of the desert regions. Marsh grapefruit is a popular cultivar that yields high quality fruit with few seeds. It is harvested in the winter and early spring.

Scion cultivars are generally budded onto rootstocks with certain characteristics that enhance fruit production. Rootstocks are selected for improving tree vigor, fruit size and quality, cold hardiness, and adaptability to soil conditions. Besides horticultural qualities, resistance or tolerance to diseases and nematodes is a major consideration.

Table 2 shows the characteristics of the main rootstocks used today in California citrus plantings. Troyer and Carrizo citrange have now largely replaced older rootstocks, such as sour orange, rough lemon, sweet orange, and Cleopatra mandarin. Sour orange is very susceptible to tristeza. Rough lemon is susceptible to cold and Phytophthora root rot, but still provides a satisfactory rootstock for lemon and grapefruit in the sandy soils of the desert regions. Sweet orange is affected by *Phytophthora* and does not provide a good stock if used for replanting in old citrus soil. Cleopatra mandarin has an average productive life and is more tolerant to salty soils than other rootstocks, but is slow in reaching full productivity.

Troyer citrange is the preferred rootstock for oranges and grapefruit in all growing areas. It grows vigorously and produces high quality fruit. Carrizo citrange, which has similar qualities, is now being planted to a greater extent, but long-term experience is still limited. Trifoliate has been used for many orange plantings in the San Joaquin and Sacramento valleys because it is the most cold-hardy rootstock. Trifoliate does not grow well on calcareous or alkaline soils and has a short productive life with certain

Table 2. Pest and Stress Susceptibility of Rootstocks Used in California Citrus Plantings[1]

	Trifoliate Orange	Troyer Citrange, Carrizo Citrange	Alemow (Citrus macrophylla)	Rough Lemon	Cleopatra Mandarin	Sour Orange	Sweet Orange
Tristeza	tolerant	tolerant	susceptible	tolerant	tolerant	very susceptible	tolerant
Phytophthora root rot	highly tolerant	tolerant	highly tolerant	very susceptible	susceptible	highly tolerant	susceptible
Armillaria root rot	susceptible	susceptible	no data	susceptible	susceptible	tolerant	susceptible
Exocortis	susceptible	tolerant	tolerant	tolerant	tolerant	tolerant	tolerant
Cold	highly tolerant	tolerant	susceptible	susceptible	tolerant	tolerant	tolerant
Citrus nematode	tolerant	less tolerant	susceptible	susceptible	susceptible	susceptible	susceptible
Salinity	susceptible	susceptible	highly tolerant	susceptible	tolerant	tolerant	susceptible
Poor Drainage	highly tolerant	susceptible	susceptible	susceptible	susceptible	tolerant	susceptible
Used with:	oranges, mandarins (possibly short-lived)	oranges mandarins (short-lived) Lisbon lemon, grapefruit	lemons only	lemon, grapefruit	oranges, mandarins, grapefruit	oranges, grapefruit	oranges, mandarins, grapefruit

[1] See text for details.

scions. Alemow (Citrus macrophylla) is currently the most popular rootstock for coastal lemon. Yields of this rootstock are high, but the productive life may be short, particularly when combined with a Eureka scion. Because of its sensitivity to cold, it is used less in interior valleys and rarely in the San Joaquin Valley. Several new rootstocks and scion cultivars are in experimental use. Check with your farm advisor for the best combination for your growing conditions.

Soil and Water Management

Citrus is grown on soils with widely differing texture and depth, but it performs best on medium to deep soil that drains well. Drainage depends on soil texture, depth, and subsoil. Sandy soils having larger, more coarse particles, and thus large pores, drain more easily than clay soils with fine particles and relatively small pores. Check your orchard site for soil depth and subsoil by digging holes in several locations; the number depends on the soil variability. Look for layers of differing soil texture and possible hardpans. Determine the soil depth up to 1.2 m (4 feet) with an auger or soil tube. If you know the depth of the soil profile, you can calculate the water-holding capacity (see Scheduling Irrigations, p. 19). Where a hardpan obstructs water movement, deep ripping, slip plowing, or explosives can improve drainage and deepen the root zone. Installing tile drains above an impervious layer has deepened the root zone in some soils with a high water table in central and southern California.

Proper water management takes into account soil texture, root depth, crop water requirements, and irrigation method. The objective is to supply the tree with the right amount of water to produce an optimum yield of high quality fruit. Keep the soil well aerated by avoiding compaction and waterlogging. Adjust your irrigation practices to the intake rate of the soil. In some cases, a ground cover helps improve the water penetration of surface soil but may also compete with the trees for available moisture. Citrus has a shallow root system; even in deep soil, more than half of the tree's feeder roots are usually in the top 2 feet of soil. Thus irrigations during the dry season have

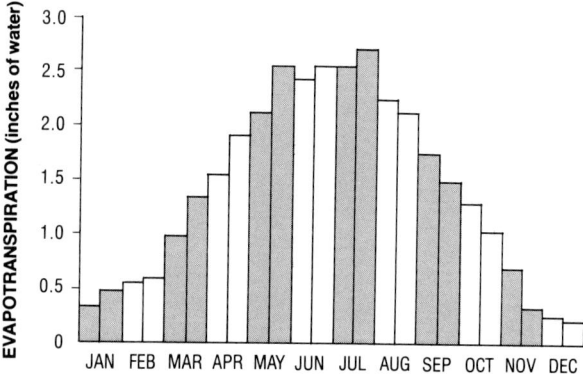

Figure 10. Evapotranspiration needs for a citrus orchard during a normal year (mature trees on bare soil in the San Joaquin Valley). Evapotranspiration measured at two-week intervals.

Figure 11. Furrow irrigation is a common practice in citrus orchards of the San Joaquin Valley. In mature orchards, three broad, permanent furrows are established between tree rows.

Figure 12. Low-volume irrigation, which is designed to meet the trees' water requirements on a continuous basis, is becoming more widely used in citriculture.

to be frequent enough to replenish the water in the top of the soil profile to prevent tree stress. In most areas, winter rains provide enough moisture into early spring. Where winter rainfall is light, as in the desert districts, additional winter irrigations may be needed. Unlike deciduous trees, citrus uses water throughout the year; water consumption is slow during the winter and peaks in June and July (Figure 10). The citrus tree's weekly water use average of 1.3 inches during the summer is lower than for deciduous trees mainly because citrus is better able to regulate water loss through the leaf pores. In California, citrus uses about 30 to 36 inches per year, with modest differences in water use requirements among mature orchards in the various growing areas.

Careful water management helps prevent certain stress and pest problems. Trees stressed for water early in the summer lose more fruit during the "June drop," the thinning of the young crop that often occurs at that time. Splitting of navel oranges is often more severe after water stress followed by high humidity. Excess water, especially in the spring, is more often a problem than lack of water. Overwatering can lead to waterlogging of the soil, contributing to root and crown diseases. Water quality is also important; excess salts in irrigation water reduce tree vigor and yield. Citrus is sensitive to high levels of boron, which cause severe leaf symptoms (see p. 123). More information on water analysis and corrective measures is available at farm advisors' offices.

Irrigation Methods. Citrus is commonly irrigated by furrows and low-head sprinklers, but low-volume irrigation systems are growing in popularity. Basin and flood irrigation is still used to some extent in the desert. The choice depends on the availability and cost of water, installation and operation costs, soil texture, slope of the land, and frost hazard. Each system can supply water adequately if managed properly.

With the furrow system, three to five furrows are made between each row of trees in mature orchards. In newly established orchards, one small furrow on either side of the tree, close to the trunk, is sufficient. In mature orchards under nontillage management, the furrows are shallow and have wide bottoms with broad ridges to support orchard traffic (Figure 11). This system limits soil compaction to the ridges and does not interfere with water penetration in the furrow bottoms. Furrows may be either level or graded. Level furrows are filled with water and allowed to stand until the water has infiltrated the soil. Level furrows can be efficient but provide poor drainage for heavy winter rains. Graded furrows usually require a tailwater return system for controlling runoff and improving irrigation efficiency.

Basin and flood irrigation is mainly used on sandy desert soils, where water penetration is fast and water demand is high. The land is leveled to near zero grade, and berms are established to form basins around one or several trees of a row. Large headstands rapidly fill these basins to the desired level, and the water is allowed to penetrate. Some systems also channel the water from one basin to the next, a labor intensive operation.

Sprinkler irrigation is suitable for all soil types and terrains, particularly for sloping or rolling land. Systems with low sprinkler heads and drag hoses are most commonly used. The amount of water discharged can be adjusted to the intake rate of the soil, thus avoiding standing water and runoff. Sprinklers are convenient for short, unscheduled irrigations—for example, when frost protection is needed. Overhead sprinklers can be advantageous for pest controls such as removing dust from foliage, but they are not widely used because they require a more expensive pumping system and high quality water to avoid phytotoxicity.

Low-volume irrigations were first developed in southern California citrus orchards to overcome high water cost and salinity problems. Many groves in the San Joaquin Valley have now switched to low volume. Low-volume systems, including drip, mist, and mini-sprinklers, are designed to meet the tree's daily water requirements. The system consists of emitters located near the trees, lateral lines, main lines, and a central control station (Figure 12). Although installation and energy costs are high, low-volume irrigation offers many advantages. It can save water and labor and allows fertilizers to be injected into the irrigation water when needed. The amount of water applied can be closely regulated by time clocks and water meters. However, low-volume irrigation systems, as well as sprinkler systems, must be well managed to maintain high efficiency.

The irrigation method chosen may affect certain pest problems. Basin or furrow irrigation can spread weeds and pathogens through the orchard as the water moves across the surface. Water from canals may be contaminated with weed seeds and pathogens. Basin irrigation promotes the development of crown diseases if tree trunks remain wet too long. Under low-volume irrigations, weed control has to be adjusted because the continuously moist zone around the emitters favors weed growth and the rapid breakdown of herbicides.

Scheduling Irrigations. To avoid over- or underirrigation, carefully time irrigations and apply the correct amount of water. If you wait until your trees show signs of stress, such as curling and wilting of leaves, fruit growth may already have been reduced. As trees mature, adjust the amount of irrigation water to account for the increased demand.

To determine how much water the trees need, it is important to understand the interaction of soil, water, and tree roots. About half of the soil volume consists of solid particles. The rest of the soil volume is pore spaces, holding varying proportions of air and water, which are vital to root function and tree health (Figure 13). The amount of water available to the trees (AW) depends on the total volume of the pore space; AW is larger for clay

Soil is about half solid material by volume (large circle). The rest of the soil volume consists of pore spaces between soil particles; pore spaces hold varying proportions of air and water (small circle).

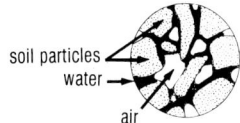

Just after a furrow or flood irrigation or heavy rain, the soil is **saturated**—the pore spaces are entirely filled with water.

After a field is allowed to drain following irrigation, the soil is at **field capacity**; in most soils, about half of the pore space is filled with water. About half of this water is **available water** that can be used by plants; the rest is unavailable because too much suction is needed to remove it from the pore spaces. The proportion of water available to plants is higher in clay soils and lower in sand.

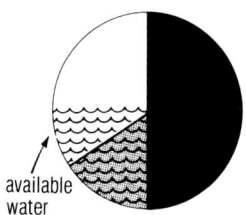

The **allowable depletion** is the proportion of the available water that can be used up before irrigation is needed. This proportion is set by the crop's tolerance for moisture stress under prevailing conditions and by the cost of supplying water.

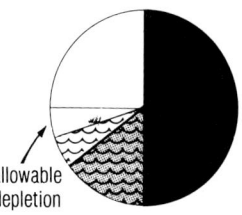

At the **wilting point**, all available water is gone; plants wilt and die unless water is added.

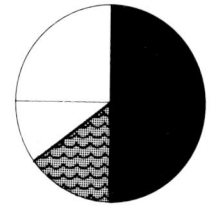

Figure 13. The soil reservoir.

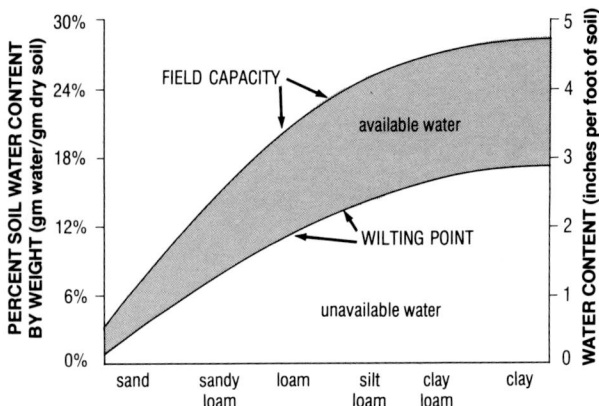

Figure 14. Available water for different soil textures. The proportion and absolute amount of water available to the plant in coarse-textured, sandy soils is less than in fine-textured, clay soils.

soils than for sandy soils, which have larger particles (Figure 14). Even though all AW is potentially available for use, it becomes harder for the roots to extract water as the wilting point (WP) is approached. Thus, the soil should be irrigated before the water level in the root zone reaches WP. The allowable depletion level depends on the plant species, its growth stage, and atmospheric conditions, so that no single recommendation can be made for all situations. For citrus, a depletion level of 50% of AW has been used successfully in many orchards. Your farm advisor or irrigation specialist can help you estimate the allowable depletion for your orchard.

You can determine when and how much water to apply by compiling a water budget. First, check your soil texture, available water (AW) at field capacity, and rooting depth. The total water holding capacity is calculated simply by multiplying AW by the depth of the root zone. For an allowable depletion level of 50%, divide AW by 2 to determine how much water can be depleted. When the soil moisture loss reaches the predetermined depletion level, it is time for an irrigation to refill the soil water reservoir.

You can estimate water loss by using evapotranspiration (ET) data. ET is defined as the amount of water lost through transpiration by the plant and through evaporation from the soil. ET tables based on long-term average conditions are available for citrus orchards with or without a ground cover and for different percentages of canopy cover. An example is given in Figure 10. Current ET data are often announced in local newspapers or on the radio, but historical data for a normal year can provide an approximation, especially in midsummer when the weather does not vary much from year to year. By adding up the weekly or biweekly ET figures until they approach the calculated depletion level, you can estimate when the next furrow or sprinkler irrigation is needed (Figure 15). To calculate the actual amount of water that needs to be applied, you must take the irrigation efficiency into account. Certain unavoidable losses occur during irrigation, mostly from deep percolation and runoff. Furrow irrigation is normally 65 to 75% efficient, whereas sprinkler and low-volume irrigation can approach 90% efficiency. Consult your farm advisor or Soil Conservation Service office for more information on irrigation efficiency. Add the calculated amount for inefficiency to the net amount of irrigation water needed at each irrigation. The amount of water you can apply at one time, however, depends on the intake rate of the soil. For example, if the soil can absorb only 2 inches within 24 hours, rather than the needed 3 inches you calculated, you have to adjust the irrigation intervals accordingly.

To confirm your calculations and verify that the soil profile has been filled, check the soil moisture after an irrigation and periodically between irrigations with soil tubes, soil augers (Figure 16) or tensiometers. Tensiometers, which are installed at different soil depths, tell how hard plants have to suck to get water from the soil (Figure 17). Quantitative evaluation of tensiometer readings is difficult because they depend on soil type, depth of instrument placement, and irrigation method. Readings can be used qualitatively, however, to check if the water has reached a certain depth after an irrigation (when suction decreases) and how dry the soil gets between irrigations (when suction increases). Tensiometers are particularly useful for low-volume irrigations because they can indicate whether or not the system is maintaining adequate moisture in the root zone. (See Leaflet 2264, *Questions and Answers about Tensiometers*, Agricultural Sciences Publications, for details on installation, reading, and maintenance of tensiometers.)

Several other instruments have been developed to make water management more accurate and easier. An instrument called the neutron probe is being used by some irrigation consultants to measure soil moisture. There is also a portable pressure chamber that measures water stress on the tree directly, using the tree itself as a tensiometer. Studies are underway to correlate the water status of the plant with that of the soil for use in irrigation scheduling. Researchers are also evaluating the infrared thermometer, an instrument that measures leaf temperature in relation to the ambient air temperature; the difference between the two can be related to plant status and irrigation requirements.

Finally, the University of California, in cooperation with the California Department of Water Resources, is researching computer programs that can accurately estimate irrigation needs in individual orchards based on crop ET requirements, weather, and specific soils information. A number of University of California publications are available on various aspects of water management. See the reference list at the end of this book.

ORCHARD/CROP _San Joaquin Valley, bare soil / citrus (mature)_

SOIL _sandy loam_

AVAILABLE WATER _1.5 in./ft. (from Fig. 13)_

ROOTING DEPTH _4 ft._

TOTAL AVAILABLE STORED WATER _1.5 in./ft. × 4 ft. = 6 in._

ALLOWABLE DEPLETION _50 %_

ALLOWABLE LEVEL OF WATER DEPLETION _6 in. × 0.50 = 3 in._

IRRIGATION EFFICIENCY _70 % (furrow)_

GROSS IRRIGATION REQUIREMENT _3 in. ÷ 0.70 ≅ 4.3 in._

DATE	ET (inches)	CUMULATIVE ET	REMARKS
Jan 1-15	0.32	0.32	assume soil at field capacity from December rains
Jan 16-31	0.46	0.78	
Feb 1-14	0.56	0.99	rained 0.5 inch Feb 2; 0.35 inch (70%) subtracted from cum. ET
Feb 15-28	0.59	1.58	
Mar 1-15	0.99	2.57	cum. ET approaching allowable depletion; irrigate soon
Mar 16-31	1.34	1.34	irrigated Mar 16; started new cum. ET
Apr 1-15	1.54	2.88	irrigate soon
Apr 16-30	1.83	1.83	irrigated Apr 18; started new cum. ET

Figure 15. Sample form for scheduling furrow or sprinkler irrigation using evapotranspiration (ET) data. Start accumulating ET January 1 at two-week intervals or more often if water demand is high. If substantial rainfall occurs after January 1, adjust ET by subtracting effective rainfall (about 70% of total) from accumulated ET. The first irrigation is needed at the end of the March 1–15 period, as the cumulative ET approaches 3.0 inches, the allowable depletion. Schedule the second irrigation for about April 18, when another 3.0 inches are about to be depleted. Apply 4.3 inches, the gross irrigation requirement, at each irrigation.

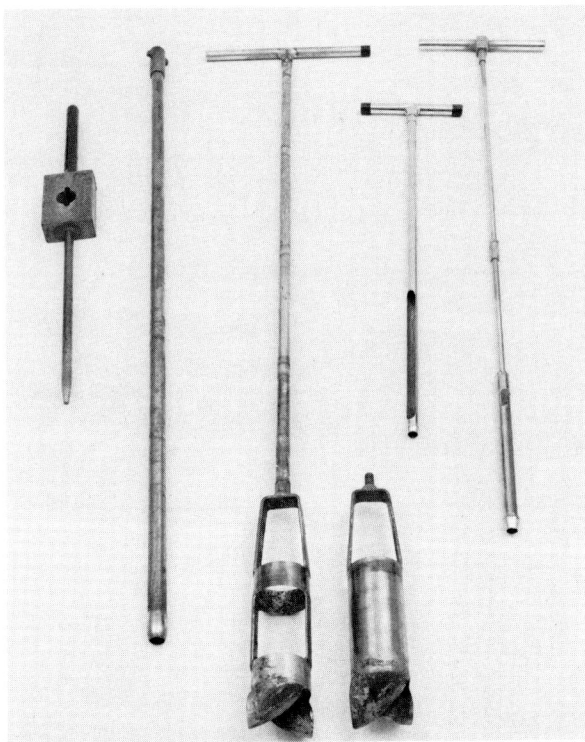

Figure 16. Tools for taking soil samples. The Veihmeyer tube (left) has a slotted slide hammer for driving the tube into the soil and removing it. Soil augers (center) are available with a variety of bits for different soil types. Oakfield soil tubes (right) are usually the easiest to use, especially for samples down to about 2 feet. These tools are useful for monitoring soil moisture. The Veihmeyer tube and soil auger are also suitable for sampling nematodes.

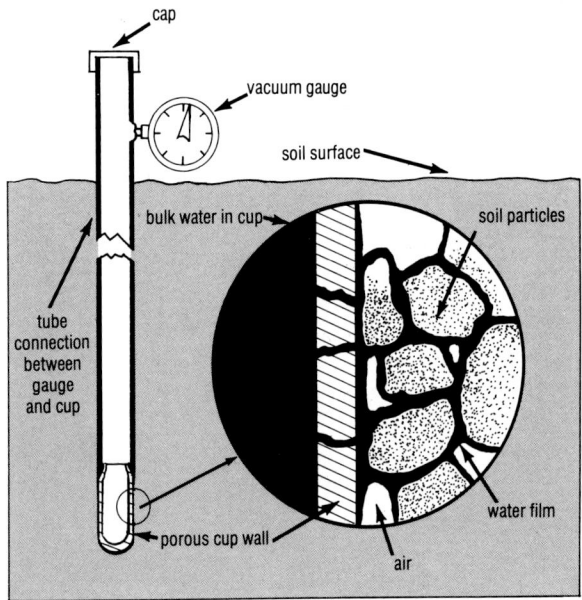

Figure 17. Principle of the tensiometer. The waterfilled cup at the bottom of the tensiometer is in contact with the water film surrounding the soil particles. As the soil dries, water is sucked out of the cup, creating a vacuum, which is measured by the gauge.

Fertilizing

Citrus trees must be supplied with adequate amounts of essential nutrients. Trees weakened by nutrient deficiencies produce less and are less able to withstand pest and frost damage than healthy trees.

Citrus orchards generally require yearly supplements of nitrogen and zinc. Often manganese is deficient, and occasionally magnesium, potassium, or iron need to be supplemented. Deficiencies of phosphorus and copper are rare. Adequate levels of nitrogen are critical in the spring for good growth and fruit set, but high levels during the summer and early fall contribute to the regreening problem of Valencia oranges and result in thicker rind and lower juice content of oranges and grapefruit. Because they set fruit in multiple crops, lemons are more tolerant of high nitrogen levels during the summer. Trees lacking sufficient amounts of the other nutrients may produce fruit of lower quality or smaller size. Deficiencies of zinc, magnesium, manganese, or iron produce typical leaf symptoms (see Mineral Deficiencies and Toxicities, p. 123).

To confirm a field diagnosis of leaf symptoms, obtain a leaf analysis from a reputable laboratory. A leaf analysis also indicates low but not yet deficient or higher than optimal nutrient levels so that you can take corrective measures before symptoms appear. Thus a leaf analysis is a valuable tool for designing a fertilization program that avoids excesses or deficiencies detrimental to fruit production and quality. Excessive application of fertilizer increases production costs, may adversely affect fruit quality, alter the absorption of other nutrients, and contribute to pollution of water supplies. The recommended time for a leaf analysis is between mid-August and mid-October. Professional laboratories analyze and also collect leaf samples. Check with your farm advisor's office for names of experienced labs.

Table 3 shows the optimum concentration range for nutrients in a leaf sample. If nutrients are deficient, they are applied either to the soil or to the foliage or both. Nitrogen can be applied to the soil, incorporated into drip irrigation water, or sprayed on the foliage. Zinc is generally applied as a foliar spray, often in combination with a pesticide treatment. Magnesium, manganese, and sometimes potassium are also sprayed on the foliage.

The ability of trees to absorb nutrients depends on many factors, including root health, soil moisture, soil acidity, presence of other minerals, and growth stage of the tree. Roots damaged by disease, nematodes, or overwatering cannot provide the tree with adequate amounts of nutrients despite an ample supply in the soil. Trees growing in alkaline soils at times develop zinc, manganese, copper, or iron deficiencies.

When a leaf analysis shows too low or too high levels of several nutrients, first correct the most severe imbalance. This may involve adding or withholding a nutrient or washing excess salts below the root zone. Retest the following year and make further adjustments as necessary.

Table 3. Leaf Analysis Guide for Nutrients Commonly Deficient or in Excess[1]

Nutrient	Optimal range[2]
nitrogen	2.4 – 2.6%
potassium	0.7 – 1.09%
magnesium	0.26 – 0.6%
zinc	25–100 ppm[3]
manganese	25–200 ppm
iron	60–120 ppm
boron	31–100 ppm
chlorine	less than 0.3%

1. Based on dry weight of tissue from 5- to 7-month-old, spring cycle leaves from nonfruiting shoots.
2. Values are for bearing navel and Valencia oranges; preliminary data suggest similar nitrogen values for lemon, slightly lower values for grapefruit. Optimal range of nitrogen is somewhat higher for young trees.
3. ppm = parts per million

Make sure you apply the appropriate fertilizer at the right time. For oranges and grapefruit, apply nitrogen to the soil or foliage in late winter or early spring; lemons may receive nitrogen at any time of the year. Foliar applications of potassium and especially magnesium are most effective when the spring growth flush is two-thirds to three-fourths expanded. Timing of zinc, manganese, and copper sprays is generally not critical, so that these nutrients can be incorporated into pesticide or growth regulator sprays. If one application of zinc does not correct the deficiency, apply the spray when the foliage is two-thirds to three-fourths expanded. If this is not sufficient, apply zinc twice, once onto the spring growth and once onto the summer growth flush. Soil applications of zinc are only effective on acid soils. When applying potassium or phosphorus to the soil, band them at the drip line of the trees; timing is not critical. Ammonia or ammonia-releasing fertilizers should be carried into the soil soon after application to avoid breakdown of the ammonia and release of nitrogen into the atmosphere. In today's non-tilled citrus orchards, rain or irrigation water serve to carry the fertilizer to the tree roots. Manure, however, is not sufficiently leached and needs to be disced into the soil. Your farm advisor can help you in setting up the right fertilization program for your orchard.

A leaf analysis may reveal toxic levels of certain minerals, such as sodium, boron, or chlorine. An excess of these minerals may be present in the soil naturally, and a soil analysis is useful in locating trouble spots. More often, however, irrigation water contains excessive levels of boron and sulfur. If the water source cannot be changed, provide good drainage, and allow for extra irrigation to wash the salts below the root zone. You may also apply soil amendments or additional nitrogen fertilizer if the leaf analysis does not indicate already high levels of nitrogen.

Frost Protection

In many citrus-growing areas of California, cold damage is a potential hazard to successful crop production. Damage symptoms on fruit are illustrated on p. 120; for symptoms on twigs, see p. 124. Two types of cold hazard exist: frost due to heat radiation, and freezes due to a cold air front moving through an area. Damage usually occurs from radiation frost when the orchard soil and lower air layers fail to retain sufficient heat from the daytime sunshine to keep the orchard frost-free through the night. Soil acts as a heat reservoir, absorbing heat during sunny days and radiating it during cold nights. If the preceding afternoon has been warm (above 16°C or 60°F), a layer of warm air remains as a relatively low ceiling above the cold ground air at night; this condition is called a temperature inversion. Under temperature inversion conditions, wind machines can mix the warm and cold layers of air, helping prevent frost damage. When a cold air front moves in, however, the ceiling is very high or nonexistent. Wind machines are largely ineffective under these conditions, and freeze damage may occur.

Frost hazard is compounded in low areas of hilly orchards when the air is dry and still. Cold air slowly flows down a slope to the lowest point; thus frost hazards are greatest on valley floors. Cold air can also accumulate in bottom lands against embankments, windbreaks, or other barriers that prevent drainage of the cold air. Dry air, which is indicated by a low dew point, allows rapid cooling because there is less moisture to store the heat. Thick fog, on the other hand, provides a thermal blanket, protecting against frost.

Certain protective measures can be taken to increase the tree's ability to withstand cold. Healthy trees well supplied with water are more able to withstand frost than weak, dry trees. Do not fertilize during late summer or early fall to avoid stimulating new growth susceptible to cold. Avoid pruning in the fall or winter as it stimulates new growth. A maximum canopy can help keep heat in the orchard. Protect young trees; at the end of October, wrap trees with insulating material, such as palm fronds, corn stalks, wood pulp, heavy corrugated cardboard, or fiberglass. Cover the trunks from the soil up to the main scaffolds; keep the leaves free. Moist soil is a better heat reservoir than dry soil; a bare, moist ground is warmer than dry soil covered with vegetation or light colored materials, such as wood chips. Where a ground cover is desirable to protect against erosion, mow it close to the ground, so that more heat absorption and radiation by the soil is possible.

Not only trees, but also tensiometers need protection. Cover them with layers of burlap or other insulating material.

Figure 18. Wind machines can often prevent frost damage by mixing cold and warm air layers.

Before the frost season starts, check your frost protection equipment. Install thermometers (see p. 14) and check them for accurate readings; the National Weather Service provides a calibration check in October, usually at county offices or citrus packinghouses. Check the operating condition of your wind machines, and know where to find a service person or spare parts in an emergency. If you use orchard heaters, store enough fuel for a few days. Make sure your heaters conform to the requirements of the California Air Resources Board.

Frost protection is a matter of economics; how much protection can you afford in order to avoid a potential crop loss? Running wind machines (one per 10 acres) may often be sufficient to prevent frost damage (Figure 18).

To use wind machines most efficiently, install a remote thermometer about 13 meters (40 feet) above the orchard where the wind machines are operating. It will tell you what the temperature difference is between the inversion layer and the orchard, so that you can determine the best strategy for operating the wind machines; this can save substantially on operating costs. If a strong inversion exists, that is, if the temperature difference is 2° to 3°F, you can delay running the machines until the temperature drops to 29° or 28°F, as the warm air will mix in quickly with the cooler air below. If you have a weak inversion (0.5°F or less), start the machines once the temperature drops to 32° or 31°F.

Heaters alone are not economical today because of high fuel costs. A combination of wind machines and heaters or running water can provide effective and affordable protection. The choice of irrigation depends on the availability and cost of water, soil drainage, and irrigation system. Basin irrigation is most effective for frost protection, as it provides a large water surface that gives off heat as it cools down to freezing. Furrows and low-head sprinklers are effective, provided a sufficient amount of water is applied. If the air is dry, however, sprinklers are more likely to cool the surrounding air as the water drops evaporate than furrow irrigation.

For all citrus-growing areas, a frost protection service operates from November through February. District meteorologists develop long-term and short-term forecasts, which are released every night over radio, television or telephone recording. You can use these forecasts in combination with your thermometer readings to predict fairly accurately the frost hazard in your orchard. For more details on frost protection, see Leaflet 2372, *Protecting Citrus from Cold Losses*, Agricultural Sciences Publications.

Harvest

Citrus fruit do not have a clearly defined point of maturity. Color is generally not a good indicator of ripeness because its appearance depends mainly on sufficiently low night temperatures. For marketing purposes, maturity standards have been established. Oranges, grapefruit, and mandarins can be harvested once they meet a minimum soluble solids to acid ratio and a minimum color standard; lemons have to reach a certain juice content. Packinghouses test representative samples and inform growers of projected harvest times.

The harvest season may extend from 4 to 9 months, depending on the cultivar and the market conditions. Navel oranges and mandarin types are harvested from November through May or even June, Valencia oranges from February through November. Coastal lemons are picked three to six times a year, mainly from March through June. Lemons in the central valley are harvested during the winter months. Coastal-intermediate grapefruit is a summer crop, picked during July and August after the harvest of the desert grapefruit is completed.

Because of the extended harvest periods, pest management actions have to be scheduled accordingly. At times, the cost of a chemical treatment can be saved if the fruit is harvested early. Where treatment is necessary, observe reentry intervals and residue tolerances. Mature fruit kept on the tree for a long time are not only exposed to damage by pests and adverse weather but will also age; fruit may drop, or the rind may develop stains, turn puffy or sticky, and become susceptible to decay. Chemicals called growth regulators are available to alleviate these problems. Growth regulators are active in minute amounts, regulating various physiological processes. Two compounds are registered for use on citrus, the isopropyl ester of 2,4-D and gibberellin A_3. For details of application and precautions, see Leaflet 2447, *Pre-Harvest Use of 2,4-D on Citrus*, or Leaflet 2903, *Treatment Guide for California Citrus Crops*, Agricultural Sciences Publications.

Pruning

Pruning is not a regular operation in citrus, with the exception of lemon. As an evergreen tree, citrus stores most of its food reserves in leaves, twigs, and branches; very little is stored in the roots. Pruning thus cuts off valuable food sources, which could be used for fruit production. Prune only when the benefits outweigh the loss in fruiting wood and the delay in production, as when training young trees, opening up a dense planting, or reducing tree height to facilitate harvest and pest control operations.

Most citrus trees do not need pruning for 2 to 3 years after transplanting to the orchard, except for occasional removing of sucker shoots on the trunk. New buds sprout from the center of the tree and from the upper side of bent branches, which droop further to form the tree's skirt. Through this repeated process, a dense canopy of upward growing and bending branches forms. From the age of 3 to 6 years, a light selective pruning may be done to remove some crowded and crossing branches.

Bearing orange and grapefruit require little pruning. At times, removing top branches or side branches is necessary. Where the root system is injured, reducing the canopy size helps in recovery. Do not open up trees; limbs that have grown in the shade are susceptible to sunburn, and an open canopy is an unfavorable environment for natural enemies.

Lemons require more pruning than other cultivars. Lemon trees produce long, upright shoots and strong laterals through the center of the tree. Selective pruning helps strengthen the shoots, prevents crowding in the tree center, and facilitates harvest.

Manual pruning is the preferred method because it allows selective removal of dead or undesirable branches close to a lateral or scaffold. To save labor costs, mechanical pruning machines for hedging or topping trees have been developed. Hedging, or trimming back the sides of the trees, is often necessary in today's close plantings. Hedging provides an economical way to open up trees, but it removes good and bad wood alike. Mechanical topping keeps trees at a manageable height. Numerous sprouts grow from the trimmed wood, forming a dense cap of foliage; occasional hand pruning is necessary to let some light back into the tree. For more details on pruning, see Leaflet 2449, *Pruning Citrus Trees*, Agricultural Sciences Publications.

Ground Cover

Maintaining a ground cover is not common practice in citrus culture. The extra costs of water, nutrients, and management and the increased frost hazard outweigh the benefits in most cases. Benefits include improving water penetration and aeration of fine-textured or compacted soil and preventing soil erosion, an important consideration on sloping land. A ground cover also reduces dust, which contributes to mite outbreaks, and it may provide food and shelter for many predators and parasites of citrus pests. More research, however, is necessary to establish benefits for pest management. A ground cover can be managed by mowing alone or by mowing the row middles and using herbicides around tree trunks. Cultivation is only practical in young orchards (see also section on weeds).

Biological Control

IPM programs in citrus depend on the many natural enemies that provide partial or complete control of various invertebrate pests. Natural enemies include parasites, predators, and pathogenic microorganisms. Some of the most effective natural enemies in citrus today were introduced from other citrus-producing countries. Native species, including parasitic wasps, lacewings, ladybeetles, syrphid flies, predaceous mites and thrips, and pathogenic fungi and viruses provide control in some cases or supplement the action of introduced beneficials. Biological control in citrus is most important for insects, mites, and snails; no reliable biological control agents for

Table 4. Biological Control of Citrus Pests Provided by Natural Enemies in the Major Growing Areas[1]

Pest	Coastal Intermediate	Interior	San Joaquin Valley	Major Natural Enemies
Amorbia	+	+	?	*Trichogramma platneri*
Bayberry whitefly	+ +	+ +	NA	*Eretmocerus* sp. (native), *Eretmocerus* n.sp., *Encarsia* n.sp.
Black anise swallowtail	+	+	+ +	*Hyposoter* sp.
Black scale	+ + +	+ + +	NA	*Metaphycus helvolus*
Broad mite	?	NA	NA	*Amblyseius stipulatus, Amblyseius hibisci*
Brown soft scale	+ + +	+ + +	+ + +	*Metaphycus luteolus*
California red scale	+ +	+ +	+	*Aphytis melinus, Aphytis lingnanensis* (mainly coast), *Comperiella bifasciata, Encarsia perniciosi* (mainly coast)
Citricola scale[3]	NA	NA	+	*Metaphycus helvolus, Metaphycus luteolus, Coccophagus lycimnia*
Citrophilus mealybug	+ + +	+ + +	NA	*Coccophagus gurneyi, Tetracnemus pretiosus*, general predators[4]
Citrus cutworm	NA	+	+	*Ophion* sp., *Banchus* sp.
Citrus looper	+ +	+ +	+ +	*Apanteles* sp., fungal and virus diseases
Citrus mealybug	+ + +	+ + +	NA	*Leptomastidea abnormis, Leptomastix dactylopii*, general predators[4]
Citrus red mite	+ +	+	+	*Euseius hibisci, Amblyseius stipulatus* (coast only), *Amblyseius limonicus* virus, *Stethorus*, other general predators[4]
Citrus thrips	+	+	?	*Euseius hibisci*, spiders, *Orius*
Citrus whitefly	+ +	NA	NA	*Prospaltella* sp., *Encarsia* n.sp.
Citrus mealybug	+ + +	+ + +	NA	*Leptomastidea abnormis, Leptomastix dactylopii*, general predators[4]
Cottonycushion scale	+ + +	+ + +	+ + +	*Rodolia cardinalis, Cryptochaetum iceryae*
Fruittree leafroller	NA	?	?	*Trichogramma* sp.
Longtailed mealybug	+ + +	+ + +	NA	*Anarhopus sydneyinsis, Tetracnemus peregrinus, Anagyrus fusciventris*, general predators[4]
Omnivorous leafroller	NA	+	+	*Erynnia tortricis, Elachertus proteoteratis, Trichogramma* sp., *Apanteles* sp.
Orange tortrix	+	+ +	NA	*Apanteles aristoteliae, Exochus* sp.
Purple scale	+ +	NA	NA	*Aphytis lepidosaphes*
Spirea, cotton aphid	+ +	+ +	+ +	*Lysiphlebus* sp., Ladybeetles, Syrphids, Lacewings
Twospotted mite	NA	NA	+ +	Sixspotted thrips, *Stethorus*, other general predators,[4] *Typhlodromus occidentalis*
Western tussock moth	NA	+	+	*Trogoderma sternale* (coast only)
Woolly whitefly	+ + +	NA	NA	*Amitus spiniferus, Cales noacki*
Yellow scale	NA	NA	+ + +	*Comperiella bifasciata* (yellow scale race)

1. Pests without known natural enemies are not listed.
2. + + + Complete biological control by native or introduced natural enemies; treatment thresholds only exceeded when upsets by chemicals or weather occur.
    + + Substantial biological control; thresholds occasionally exceeded.
    + Partial biological control; thresholds often exceeded.
    ? Extent of biological control not known
    NA Not applicable; pest rare or absent.
3. Generally kept at low levels by insecticides used for California red scale.
4. General predators that feed on one or several of the pest species include: lacewings, mealybug destroyer, dustywings, *Sympherobius*, *Scymnus*, *Lindorus* (prefers armored scales).

plant pathogens, weeds, or vertebrates are currently known.

Over the past hundred years, several important citrus pests have been brought under biological control by introduced natural enemies. One of the earliest and most successful examples was the establishment of the predaceous vedalia beetle, *Rodolia cardinalis*, and a parasitic fly, *Cryptochaetum iceryae*, against the cottonycushion scale. Subsequently several other major pests were controlled biologically, including the black scale, brown soft scale, purple scale, yellow scale, mealybugs, citrus whitefly, and woolly whitefly. Table 4 shows the natural enemies and the degree of control they provide when undisturbed by chemical treatments or adverse weather. In general, natural enemies provide better control in southern California than in the San Joaquin Valley, although the degree of control varies from good in coastal areas and fair in interior districts to poor in the desert regions.

Several parasitic wasps today partially or completely control the California red scale in many areas. *Aphytis melinus* is the dominant parasite species in many coastal-intermediate and interior areas of southern California. In districts close to the coast, *Aphytis lingnanensis* and *Encarsia perniciosi* are most prevalent. *Comperiella bifasciata*, an effective parasite of the yellow scale, has a host-specific race that supplements the control of the California red scale. In the San Joaquin Valley, *Aphytis* and *Comperiella* are established but do not provide economic control of the California red scale in commercial citrus groves. The native predaceous mite, *Euseius* (*Amblyseius*) *hibisci*, a viral disease, and general predators play an important part in regulating the populations of the citrus red mite. *E. hibisci* also shows promise as a predator of the citrus thrips. In recent years, a predaceous snail, *Rumina decollata*, has provided substantial control of the brown garden snail in some orchards in southern California. (For detailed descriptions and photographs of the natural enemies, see the individual pest sections.)

The search for natural enemies of established pests and new invaders continues. Efforts are underway to find effective biological control for California red scale in the San Joaquin Valley. Several parasites have recently been released against the bayberry whitefly, a new and potentially serious pest of citrus; the effectiveness of these parasites is currently under investigation.

You can enhance and supplement the action of natural enemies established in your growing area in a number of ways. In southern California, ants often hamper the activity of parasites and predators. Ants feed on honeydew and protect the honeydew-producing insects, such as soft scales, mealybugs, and aphids, from their natural enemies. They also interfere with the biological control of pests that do not produce honeydew, such as California red scale and citrus red mite. The ants have no effective natural enemies, and chemical treatment of their ground nests is often needed to preserve the biological control of the citrus pests (see Ants, p. 99).

When a foliar application for an insect or mite pest becomes necessary, choose selective chemicals whenever possible or avoid repeated application of nonselective materials; disruption of biological control leads to additional treatments and additional costs. Selective chemicals least disruptive to parasites, lacewings, and ladybeetles are generally those that are also least toxic to honeybees (see Leaflet 2903, *Treatment Guide for California Citrus Crops*, Agricultural Sciences Publications). Table 5 lists most insecticides and miticides available in citrus and their effect on various natural enemies.

Dusty conditions in and around the orchard hamper the activity of natural enemies. To keep dust to a minimum, oil orchard roads, drive slowly, and reduce traffic through the grove.

If adverse weather or an insecticide treatment has disrupted the biological control of a pest, you can hasten reestablishment by releasing additional parasites or predators. Several cooperative insectaries and private companies rear natural enemies such as *Aphytis*, *Metaphycus*, *Cryptolaemus*, and *Rumina decollata*. Check with your farm advisor for details about obtaining natural enemies for release.

Pesticides

Within an IPM program, pesticides are valuable pest management tools because they can reduce pest populations drastically within a short time. In many cases, they are the only control tools available.

Pesticides, including insecticides, herbicides, fungicides, rodenticides, and nematocides, belong to a variety of chemical groups with very different properties. Often you can choose from several chemical classes and different formulations to control one pest. On the other hand, you may be able to select one chemical to control several pests at the same time or use compatible materials together. The choice will depend on the degree of control needed, type of application, weather conditions, effect on other pests and beneficial species, and on economic and legal restrictions. When applying a chemical to foliage, choose the least disruptive chemical possible. For example, synthetic organic materials, such as organophosphates and carbamates used for control of the California red scale, are more likely to kill natural enemies than refined petroleum oils.

Oils are valuable insecticidal materials in citrus because they have little residual toxicity for beneficial insects and effectively control a number of scale and mite pests. Oil sprays, however, require certain precautions. Soil moisture, weather conditions, growth stage of the fruit, and cultivar have to be considered to avoid injury (see p. 27). You can choose from several grades of oil classified according to the size of their hydrocarbon molecules. The molecule size, which determines volatility and thus persistence, increases from narrow-range (NR) 415 to light-medium, medium, and NR 440 oil. Persistence should be

Table 5. Insecticides and Acaricides Commonly Used in Citrus and Their Effect on Natural Enemies.

Material and Target Pest	Natural Enemies	Toxicity to Natural Enemies[1]
Acephate (ORTHENE®) citrus thrips	most predators and parasites	high
Azinphosmethyl (GUTHION®) orangeworms	convergent ladybeetle *Euseius hibisci*	high moderate to high
Bacillus thuringiensis fruittree leafroller California orangedog	most predators and parasites	no or very low
Carbaryl (SEVIN®) orangeworms, scale insects	predaceous mites (SJV) Hymenoptera	no or very low high
Chlorobenzilate (ACARABEN®) bud mite, rust mite	parasites, predaceous insects predaceous mites	no to low high
Chlorpyrifos (LORSBAN®) broad mite, orangeworms	predators or parasites	no data
Cryolite (KRYOCIDE®) orangeworms	parasitic Hymenoptera predaceous mites green lacewing tachinid fly convergent ladybeetle	no to moderate no or very low no to moderate moderate no to moderate
Cyhexatin (PLICTRAN®) citrus red mite, spider mites	*Aphytis melinus* *Euseius hibisci*	low low
Dicofol (KELTHANE®) spider mites, broad mite	parasitic Hymenoptera predaceous mites green lacewing coccinellid ladybeetles sixspotted thrips	no to moderate high no or very low no to low no or very low
Dimethoate (DIMETHOATE®) citrus thrips	most predators and parasites	high
Fenbutatin-Oxide (VENDEX©) citrus red mite, spider mites	*Amphytis melinus* *Amblyseius hibisci*	low low
Formetanate hydrochloride (CARZOL®) citrus thrips	Hymenoptera	moderate to high
Malathion scale insects	most predators and parasites	high
Methidathion (SUPRACIDE®) California red scale	most predators and parasites	high
Methomyl (LANNATE®, NUDRIN®) orangeworms	most predators and parasites	highly toxic but short residual activity
Mevinphos (PHOSDRIN®) orangeworms	parasitic Hymenoptera green lacewing convergent ladybeetle predaceous mites	moderate to high moderate to high high moderate
Naled (DIBROM®) katydids, orangeworms	parasitic Hymenoptera green lacewing convergent ladybeetle predaceous mites	moderate to high no to moderate high moderate
Oxythioquinox (MORESTAN®) citrus red mite, spider mites	*Aphytis melinus* predaceous mites	low to moderate moderate
Parathion scale insects, orangeworms	predaceous mites Hymenoptera convergent ladybeetle	low moderate high

Material and Target Pest	Natural Enemies	Toxicity to Natural Enemies[1]
Propargite (OMITE®) citrus red mite, spider mites	*Euseius hibisci* *Aphytis* sp. *Comperiella bifasciata* sixspotted thrips	low low low no or very low
Ryania citrus thrips	most predators and parasites	no to low
Sabadilla citrus thrips	most predators and parasites	no to low
Spray oils-low volume (NR 440 and NR 415) citrus red mite, California red scale, other mites and scales	parasites, predaceous mites	no to low
Spray oils-high volume (NR 440 and NR 415) citrus red mite, California red scale, other mites and scales	parasites, predaceous mites	moderate
Sulfur broad mite, rust mite, flat mite	parasitic Hymenoptera predaceous mites green lacewing damsel, big-eyed, minute pirate bugs tachinid fly convergent ladybeetle	low to high moderate to high no to low low to moderate no to low no to low
Trichlorfon (DYLOX®) orangeworms	Hymenoptera green lacewing convergent ladybeetle syrphid flies	low no to low low to high high

1. Information is based on research and long field experience.

long enough to kill the pest, yet short enough not to affect the trees. Your choice will depend also on the temperature and the amount of oil you want to apply. At higher temperatures, oil dissipates faster than at lower temperatures. Thus, an NR 415 spray is suitable for the cooler fall and spring, whereas an NR 440 gives longer control during the warmer periods. Low-volume sprays, which use much less oil than dilute sprays, reduce persistence and damage potential. Where the damage potential is too great for a regular oil spray, oil can be added at a low percentage to certain synthetic organic chemicals to improve coverage. For details of application precautions, see the latest Leaflet 2903, *Treatment Guide for California Citrus Crops*, Agricultural Sciences Publications.

Among the most selective materials available are preparations of disease-causing microorganisms. These microbial insecticides are desirable control tools in IPM programs because they act selectively against certain insect groups and are nontoxic to natural enemies and nontarget species. *Bacillus thuringiensis*, for example, is effective against certain orangeworms.

Treat for insects and mites only when monitoring indicates that damaging populations are present or anticipated. Choose the appropriate pesticide and apply it when the target pest is in its most vulnerable stage; for instance, synthetic organic materials are most effective on the California red scale in late spring before the crawlers move out onto fruit and form the scale cover; in late summer and early fall, oils are the preferred materials.

Using the correct dosage, choosing the appropriate application technique and equipment, and calibrating the machinery properly are necessary for good pest control. The recommended dosages and frequency of application given on pesticide labels may be higher than necessary for good control when properly timed and applied. Using lower rates, however, requires thorough knowledge of local conditions. Recommended dosages for herbicides usually have to be adjusted for soil type, climatic conditions, and irrigation method used. Consult your farm advisor for the lowest effective dosage and best application method for the pest problem in your orchard. Before choosing a pesticide, check the latest Leaflet 2903, *Treatment Guide for California Citrus Crops*, Agricultural Sciences Publications. Always read the pesticide label; it outlines safety precautions, legal requirements, and registered uses for each product.

Pesticide Application. The success of many IPM programs depends on the correct application of the needed pesticides. The following discussion focuses on the application of insecticides; see the weed section for details on herbicide application and University of California Leaflet 2710 for calibration of herbicide sprayers.

The nondeciduous citrus tree is one of the most difficult spray targets. Healthy, mature trees are on the average 5 to 6 m (15 to 20 feet) and can be up to 8 m (26 feet) tall, and they have a dense canopy with many leaves, twigs, fruit, and branches. In addition, the pests themselves, with their different biologies and habitats, have to be considered. For achieving the most efficient pest control, choose the correct coverage, ground speed, gallonage, and type of equipment.

Coverage. Coverage—the distribution of a spray on or within the tree—depends primarily on droplet size, amount of spray, and ground speed (Table 6). The degree of coverage required differs with the target pest. For example, scale insects, which are distributed throughout the tree and do not move around much, require a spray that reaches both exterior and interior parts of the tree. Therefore, scales are most often controlled with a thorough coverage (TC) application, which uses a high gallonage and large droplets (dilute spray) and leaves a wet

Table 6. Description of Various Types of Coverage[1]

Type of Coverage	Description of Coverage	Basis for Dosage	Basis for Spray Volume	Suitable Equipment	Optimal Ground Speed (mph)
TC (thorough coverage)	high gallonage applications, large droplets, achieving thorough film wetting of all interior parts of the tree	tree	see Table 7	oscillating boom	1.4
				air blast machine with air tower	1.4
			film wetting by eye	hand spray rig with tower	not applicable
TDC (thorough distribution coverage)	median gallonage applications, large droplets, achieving thorough distribution to all interior and exterior parts of the tree without the necessity of obtaining film wetting	tree	see Table 7	oscillating boom	1.4
				air blast machine with air tower	1.4
			by eye	hand spray rig with tower	not applicable
OC (outside coverage)	median gallonage application, large droplets, achieving thorough distribution to outside or peripheral parts of the tree only	acre	500 gal, or less, per acre	oscillating boom	1.5 to 3.0
				air blast machine	1.5 to 3.0
MS (mist spray)	low gallonage applications, varying droplet size, achieving limited droplet depositions on tree surface	acre	100 to 300 gal per acre	air blast machine	varies with pest
				low volume machine	1.5 to 3.0
LV (low volume)	low gallonage, small droplets, achieving droplet deposition on all interior and exterior parts of the tree	acre	100 gal, or less, per acre	low volume machine with air tower	1.4 (red scale)
				air blast machine with low volume manifold and air tower	1.5 to 2.0 (red mite)

1. See text for details. Further information and treatment recommendations are given in Leaflet 2903, *Treatment Guide for California Citrus Crops*, Agricultural Sciences Publications.

film of spray on all tree parts. Thorough distribution coverage (TDC) sprays, which are similar to TC but use 5 to 10 gallons less spray per tree, can provide adequate thorough coverage with some materials, although they do not achieve complete film wetting. An outside coverage (OC) spray provides control for those pests that feed mainly on the outside canopy, such as citrus thrips, aphids, and certain orangeworms. OC is achieved with medium gallonage and large droplets. A mist spray (MS), which uses low gallonage and varying droplet sizes, distributes the material throughout the trees but achieves only limited coverage. MS is sufficient for active pests, such as citrus thrips, which move about and come into contact with widely spaced droplets. MS and OC coverage save on application material and refill time while they can be applied with the same machinery as TC and TDC, primarily with air blast sprayers.

Low-volume (LV) sprays provide reasonably good coverage of exterior and interior parts of the tree. They can be used for control of mites and citrus thrips, as well as for scale control with certain insecticides. The basis of a low-volume application is the small droplet size of the spray. As illustrated in Figure 19, a given amount of spray distributed in small droplets by a low-volume sprayer covers a much larger area than the same amount applied in large droplets in a dilute spray. The small droplets adhere well to the tree parts, and there is virtually no drip or runoff, as generally occurs with dilute sprays.

Ground Speed. An important variable affecting coverage is the travel speed at which you operate your sprayer. If you drive too fast, much of the spray will not reach the target; if you drive too slowly, the spray will overlap and be wasted. TC and TDC require a fairly low ground speed (1.4 mph), whereas OC and MS coverage can be achieved at 1.5 to 3 mph. The travel speed for LV depends on the target pest; it is lower for California red scale (1.4 mph) than for citrus red mite (1.5 to 2 mph) (Table 6).

Gallonage. Gallonage refers to the amount of spray mixture applied per acre or per tree. For TC and TDC, tree height and density of foliage determine gallonage per tree. Guidelines for these adjustments are given in Table 7. Guidelines for oranges and grapefruit are based on visibility into the tree when the surface foliage is held open. Because of their different shapes and pruning, only an average density is given for lemons. The numbers in Table 7 are guidelines only; adjustment in the spray amount may be needed to compensate for orchard conditions. Required gallonage varies with application method and required coverage.

For MS, OC, and LV coverage, dosage is calculated per acre. The machines are set up to deliver the recommended gallons per acre, generally without adjustment for size and density of the trees.

Figure 19. Coverage depends largely on the droplet size of the spray. The drawing shows that the same amount of spray from a concentrate sprayer can cover a much larger area than that from a dilute sprayer. For each 300-micron droplet formed by a dilute sprayer, a concentrate sprayer can discharge 6x6x6 = 216 droplets of 50 microns. (As illustrated, each droplet is surrounded by a 100-micron zone, in which each spray droplet is effective.)

Figure 20. The oscillating boom sprayer ejects a pesticide/water mixture through guns oscillating on a vertical boom. Boom sprayers are used for high volume (dilute) applications.

Figure 21. The air blast sprayer uses a large air volume to carry the spray into the tree. With the proper adjustments, this machine can apply a wide range of gallonages.

For dosages of active ingredients, consult the pesticide label or Leaflet 2903, *Treatment Guide for California Citrus Crops*, Agricultural Sciences Publications.

Table 7. Guidelines for Calculating Gallonage per Tree for Thorough Coverage[1]

	GALLONS PER FOOT OF TREE HEIGHT	
Canopy density[2]	Oscillating boom	Air blast sprayer
Orange and Grapefruit		
1. Very dense (tree trunk cannot be seen)	1.6	1.3
2. Dense (tree trunk can be seen)	1.5	1.2
3. Average density (inside foliage shell beyond trunk can be seen)	1.4	1.1
4. Thin (can see through and beyond tree)	1.1	.9
Lemon		
Average density	1.2	1.0

1. For thorough distribution coverage, use 5–10 gallons less per tree, depending on canopy density.
2. Hold surface foliage open to check canopy density.

Application Equipment. The various spray machines operate on different principles to form the spray droplets and propel them into the trees. The oscillating boom sprayer uses hydraulic pressure to move the pesticide mixture through guns oscillating on a vertical boom (Figure 20). The optimal spacing of the guns on the boom is 35 cm (14 inches). The guns should have a two-hole whirl plate, a flat, single-orifice disc, and a longitudinal central rod, which can be moved forward or backward to adjust the spray jet. To achieve good coverage in the treetops, place the discs with the largest orifices (9/64 inch or 10/64 inch) at the top of the boom; use medium-sized (8/64 inch or 7/64 inch) for the middle guns and small (6/64 inch) for bottom gun (see sample arrangement, p. 34). Flare the spray angle of the bottom gun as needed to cover the tree skirt; adjust all other guns to provide the optimal spray cone angle of 16 degrees. The optimal ground speed is 1.4 mph, but a range of 1.25 to 1.5 is acceptable. A pump pressure of 450 to 550 psi and an oscillation rate of 62 to 70 oscillations per minute are critical for good coverage.

An air blast sprayer uses moderate pressure to form and discharge spray from atomizing nozzles on a manifold into a high velocity, large air volume to carry and distribute the spray into the tree (Figure 21). Air blast sprayers without an air tower cannot achieve thorough coverage of trees taller than about 3.4 m (11 feet); module sections are available to adjust the tower for different tree heights. You can apply a wide range of gallonages with air blast sprayers, and you can use certain models for low-volume spraying by adjusting pump pressure, air speed (rpm), and nozzling.

Low-volume sprayers eject a very fine spray into the tree (Figure 22). Under low pump pressure, the spray mixture passes through orifices with diameters of 3/16 to 5/16 inch onto the surfaces of plates positioned in a high velocity air stream (greater than 175 mph). The air passing over the plate at high speed shears the liquid into fine droplets of 50 to 150 microns, then deposits the droplets onto surfaces of fruit, foliage, and branches at the interior and exterior of the tree. Depending on the label, you can apply spray oils, insecticides, acaricides, fungicides, nutrients, and growth regulators as low volume sprays. Low-volume sprayers have the advantage of covering more acreage without a refill; for example, one 500-gallon tank can spray 5 acres at low volume instead of only one-sixth acre at 3000 gal/acre dilute spraying. Just as for air blast sprayers, an air tower is necessary for low-volume sprayers to achieve satisfactory coverage in the tops of mature trees.

Low-volume spraying requires great precaution because the chemical concentration per volume unit is much higher than in dilute spraying. In addition, longer residual activity of certain pesticides has been observed; carefully follow the instructions on the label concerning reentry and harvest intervals.

Hand spraying is today used only for spot treatments, dense plantings, and trees next to buildings, borders, or on hilly terrain. Most commonly a dilute spray is applied with a pump pressure of 550 psi and a disc orifice size in the gun of 8/64 inch. For large trees, a minimum crew of one applicator on the ground and one on a hydraulic tower, mounted on the spray rig, is needed. In a widely used routine, the ground applicator first enters the tree or holds the canopy open and sprays the inside with a flared spray stream in horizontal movements, zigzagging from the top to the bottom and wetting all visible interior surfaces. He repeats the procedure in each quadrant, then moves away from the tree, adjusts the gun to a narrow, solid stream, and makes his way around the tree while spraying the outside of each quadrant in a consistent pattern. He starts by spraying the tree surface in front of him from top to bottom with horizontal strokes in a zigzag pattern, then from top to bottom with vertical strokes moving from as far as he can reach behind to the far edge ahead. While he moves to the next quadrant, he keeps the gun moving in vertical strokes to avoid wasting spray. This procedure, together with a second applicator on the tower spraying into the treetop, achieves thorough coverage of all foliar and fruit surfaces of the tree.

Aircraft spraying is an option for certain outside coverage treatments. At times, air spraying is also used if the terrain or ground conditions do not allow ground application. Spraying by aircraft is a special technique carried out only by licensed pest control operators. Drift is a major problem and thus aircraft spraying should be reserved for specially needed treatments.

Figure 22. The low-volume sprayer produces a very fine spray. It can achieve good coverage with a low spray volume because it forms much smaller droplets than an air blast or oscillating boom sprayer.

a. **List the known criteria of your operation:**

1. row spacing (trunk to trunk, ft.) _____

2. tree spacing (trunk to trunk, ft.) _____

3. tree height (ft.) _____

4. diameter of tree skirt _____

5. gal. of spray per ft. of tree height
(example from Table 7: density category 3, oscillating boom) _____

6. ground speed (mph), from Table 6 _____

b. **To set up the nozzling of the sprayer, you need to know the gal. of spray required per min. (gpm) for one side:**

$$gpm = \frac{gal./ft. \text{ of travel}}{min./ft. \text{ of travel}}$$

1. To find gal./ft. of travel, multiply the gal./ft. of tree height by the tree height: _____

   Then divide by 2 to get gallonage per one side: _____

   Then divide by the diameter of the tree skirt: _____

2. To find min./ft. of travel, use

$$\frac{60 \text{ min.}}{mph \times ft./mi.} = \underline{} = \underline{}$$

3. Then gpm for one side is: _____

Approximately 86 gal. per min. have to be apportioned among the nozzles on one side.

c. **gal. per A. (gpa)**

gpa = gal./ft. travel × tree spacing × no. of trees/A. × 2 sides/tree

Everything needed in the equation is known except no. of trees/A.

$$\text{no. of trees/A.} = \frac{sq. ft./A.}{tree \text{ spacing} \times row \text{ spacing}} = \underline{} = \underline{}$$

Now you can calculate the gpa for the example:

gpa = _____ = _____

So it takes about 2772 gal. per A. to deliver 22.4 gal. per tree at a ground speed of 1.4 mph. This calculation accounts for spray delivered between the rows as the machine moves through the orchard, as well as the 22.4 gal. delivered to each tree.

Figure 23. Worksheet for calibrating orchard sprayers for thorough coverage.

For all equipment, proper maintenance and calibration is critical for achieving good pest control. Check nozzles regularly for plugging and wear, and replace those that are worn; metallic discs wear faster than ceramic ones. Frequent use of wettable powders wears out nozzles faster than use of emulsifiable concentrates or other liquids. Dilute sprayers need more frequent maintenance of hoses and high pressure pumps.

Calibration. Effective coverage and control depends on accurate calibration, that is, correct adjustment of the sprayer so that it delivers the recommended gallonage. To calibrate any sprayer you must first determine how many gallons per minute (gpm) must be discharged to deliver the proper amount of material to each tree or acre. Then you must equip your machinery with nozzle sizes and arrangements that will emit that amount of spray when the machinery is moving at the required ground speed. The calibration steps differ according to whether trees or acres are the basis for determining gallonage. Figure 23 gives the steps for calibrating an oscillating boom or airblast sprayer for thorough coverage (or thorough distribution coverage). Use the calculated gallons per tree to determine the gallons per minute (gpm) spray output needed for one side (only one side is generally operated because of limited pump capacity). The calculated gpm has to be divided among the nozzles of one side.

For the oscillating boom, the gpm for different orifice sizes depends on the type of gun and other variables, so each boom should be calibrated. Determine the gpm for different orifice sizes by spraying a known amount of water with, for example, 10 nozzles of a certain orifice size at a certain pressure (make sure high tank volume and pressure are maintained during the time interval). Carry out the procedure for 5 different orifice sizes (6/64 to 10/64 inch) and 3 different pressures (450, 500, 550 psi). Then you can choose the correct nozzle sizes to deliver your calculated total gpm. As an example, for the 16-foot trees in Figure 23, use 14 guns on the boom, with a pump pressure of 525 psi and the following arrangement of orifice sizes (multiply each by 1/64 inch):

BOTTOM 6 7 7 7 7 7 8 8 9 10 10 9 7 TOP

Air blast sprayers use nozzle types different from those on the oscillating boom. Manufacturers provide nozzle specifications and tables with the gpm output for various orifice sizes and pump pressures. Use the tables to choose the appropriate orifice sizes to deliver your calculated gpm. Clogging can be a problem with a large number of small orifice discs, so try to use a smaller number of larger-diameter orifices. Use large diameter orifices at the top where the spray is directed to the inside top of the trees. Depending on the pump output capacity of your machine, you may only be able to run one side of the sprayer.

From the gpm you can determine the total gallons you need per acre. Then you can calculate how much active ingredient you have to mix into the total gallonage to apply the pesticide at the dosage recommended on the label or in the Leaflet 2903, *Treatment Guide for California Citrus Crops*, Agricultural Sciences Publications. Leaflet 2718, *How Much Chemical Do You Put in the Tank*, Agricultural Sciences Publications, discusses this step in detail.

To calibrate sprayers for mist spray, outside coverage, or low-volume applications, use the steps provided in Figure 24. The recommended gallons per acre and ground speed (from Table 6 or Leaflet 2903, *Treatment Guide for California Citrus Crops*, Agricultural Sciences Publications), and the row spacing provide the bases for calculating the needed gallons per minute for one side. Choose the nozzling according to the procedures provided for your sprayer. If machines have air towers or volutes, all discharge outlets must be included in nozzle output calculations. For low-volume machines, the table supplied for calibration is based on water. Oil spray mixtures require adjustment of the metering control; a guideline is to multiply the calculated gpm by 1.26. You can check the spray output by disconnecting the tubing for one set of nozzles at the manifold and collecting the spray into a container for a 1-minute interval.

To check the ground speed for all equipment, follow the steps in Figure 25. Multiplying the minutes per foot of travel by a distance in the orchard gives you the time it takes to pass that distance. If it takes you more or less time than calculated, adjust your speed accordingly. For all applications, check the distribution of the spray. After you have sprayed three rows, inspect the center row. Look for dry areas on the branches, foliage, and trunk that have been missed by the spray. For TC, check also the top center of trees. If the coverage is inadequate, rearrange the nozzles accordingly. Once you have adjusted your sprayer, find out how much pesticide you need per tank to apply it at the recommended rate. For details, see Leaflet 2718, *How Much Chemical Do You Put in the Tank*, Agricultural Sciences Publications.

Problems Associated With Pesticide Use. The attributes that make pesticides valuable control tools may in some circumstances bring problems that directly influence the success of an IPM program: the development of pesticide resistance, pest resurgence, or secondary pest outbreaks. Pesticide applications may also affect honeybees, neighboring crops, wildlife, and human health.

Pest Resurgence and Secondary Pest Outbreak. Pesticides that destroy or disturb natural enemies as well as kill a target pest sometimes cause pest resurgence or secondary pest outbreaks. These side effects are most often associated with the use of insecticides and miticides.

Pest resurgence occurs when a pesticide destroys the natural enemies of a target pest. Because the natural enemies depend on the pest for food, they take much longer

---

To find the gal. per min. (gpm) for the nozzle manifold (one side), use the following equation:

$$gpm = \frac{gpa \times mph \times row\ spacing}{1000}$$

1. ground speed (from Table 7 or treatment guide) _____

2. recommended gal. per A. (gpa) (from Table 7 or treatment guide) _____

3. row spacing (in ft.) _____

$$gpm = \frac{\_\_\_ \times \_\_\_ \times \_\_\_}{_____}$$

3.3 gal. per min. need to be discharged from one side (6.6 gpm from both sides) to deliver 100 gal. per A. at 1.5 mph.

---

1. The equation is derived from:

$$gpm = \frac{gal./A.}{2\ sides} \div min./A.$$

$$= \frac{gpa}{2} \div \frac{sq.\ ft./A.}{ft./min. \times row\ spacing}$$

$$= \frac{gpa}{2} \times \frac{ft./min. \times row\ spacing}{sq.\ ft./A}$$

$$= \frac{gpa}{2} \times \frac{mph \times ft./mi.}{60\ min.} \times \frac{row\ spacing}{sq.\ ft./A.}$$

$$= \frac{gpa}{2} \times \frac{mph \times 5280}{60} \times \frac{row\ spacing}{43560}$$

$$= \frac{gpa \times mph \times 88 \times row\ spacing}{2 \times 43.56 \times 1000}$$

$$= \frac{gpa \times mph \times row\ spacing}{1000}$$

Figure 24. Worksheet for calibrating orchard sprayers for outside coverage, mist sprays, or low volume application.

---

a. Calculate the min. per ft. of travel

(same as Fig. 23, b.,2).

$$\frac{60\ min.}{mph \times ft./mi.} = \frac{\_\_\_}{\_\_\_ \times \_\_\_} = _____$$

b. Multiply min./ft. travel by a distance in the orchard, for example, 5 row spacings (each 22 ft.)

= _____

= _____

If it takes you more or less than 53 seconds to pass 5 tree rows, adjust your ground speed.

Figure 25. Worksheet for checking ground speed.

than the pest to build to their former numbers. Meanwhile, pest individuals that survive the treatment breed without the restraint of natural enemies, sometimes building to greater numbers than existed before the treatment. In citrus, resurgence of mites may occur after a treatment with a nonselective chemical.

A secondary outbreak can occur when a pesticide destroys the natural enemies of a pest that was not the target of the application (Figure 26). Released from the pressure imposed by its enemies, this second species may then increase to damaging numbers. For example, treatments for citrus thrips or orangeworms can cause outbreaks of the citrus red mite. To reduce the chances for pest resurgence and secondary pest outbreaks, choose selective materials when possible (Table 5) and time applications to be most destructive to the pest species but least destructive to the beneficial species.

A problem similar to secondary outbreak can occur when herbicide applications kill most of the weed species present but allow a few tolerant species to survive. With the competing weeds removed, the tolerant species grow more abundantly. In many citrus orchards, for example, the repeated use of simazine has eliminated most broadleaf annuals, and resistant grasses have increased; several broadleaf species are resistant to the commonly used herbicide diuron.

Pesticide Resistance. Some populations of certain pests, mostly insects, mites, and pathogens, have developed the ability to survive application of pesticides at rates that once killed most individuals of that species (Figure 27). Although less common, resistance has also been reported in some rodent and weed species. Once resistance develops, the pesticide involved is no longer useful. Switching to another pesticide or developing a new pesticide does not necessarily provide a solution because species resistant to one pesticide often develop "cross-resistance" to others, even to materials that are not in the same chemical class as the first pesticide. Resistance to new pesticides then shows up much more rapidly than it did with the original pesticide.

Historically, the development of resistance has been an important factor in citrus pest control. The first reported case was the resistance of the California red scale to the hydrogen cyanide fumigant. In recent years, citrus thrips have shown tolerance to some organophosphates. To slow the development of resistance, use pesticides only when needed and apply the materials correctly to achieve optimal control with one treatment.

Hazards to Honeybees. Most citrus cultivars, except tangerine and tangelo, do not require pollination, but honeybees forage extensively in the orchards during bloom (Figure 28). As one of the early-blooming crops, citrus provides a valuable food source for honeybees as they build their colonies. Strong colonies are desirable

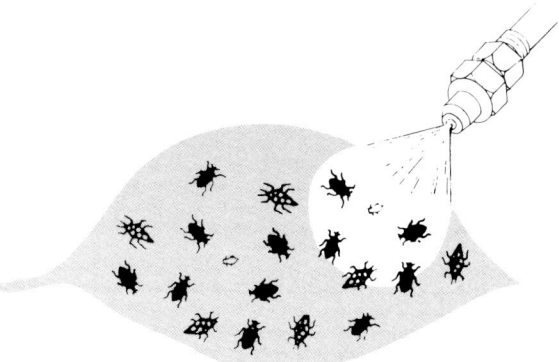

A pesticide applied to control pest A also kills natural enemies that are controlling pest B.

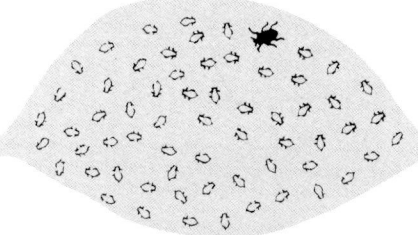

Released from the control exerted by natural enemies, pest B builds up to economically damaging levels

 pest A

pest B

 natural enemy

Figure 26. The destruction of natural enemies often results in secondary outbreaks of insects and mites.

later on to pollinate a number of summer crops, such as melon, squash, and seed alfalfa, and the following year's almond and fruit trees. Citrus contributes about 25% to the honey production of the state.

Many insecticides used in citrus are toxic to honeybees, killing bees directly, while they forage in the orchard, or indirectly, through contaminated food brought back to the hive. Insecticides have been classified according to their toxicity, ranging from highly toxic (Group I) and moderately toxic (Group II) to relatively nontoxic materials (Group III). Group III chemicals can be applied without restrictions except those provided on the label. Group II materials require notification of the beekeeper and should be used with caution. The use of Group I materials is not allowed during bloom without a special permit. You can find the toxicity categories in the Leaflet 2903, *Treatment Guide for California Citrus Crops*, or in Leaflet 2883, *Reducing Pesticide Hazards to Honey Bees*, Agricultural Sciences Publications.

Bee protection in California is enforced by county agricultural commissioners. In the San Joaquin Valley, the California Department of Food and Agriculture has designated a citrus/bee protection district. It includes the areas within 1 mile of any citrus planting of 0.4 hectare (1 acre) or more in the counties of Tulare, Fresno, and Kern. During the period of 10% bloom to 75% petal fall, as declared for several subdistricts by the agricultural commissioners, the use of Group I and II materials is under special regulation.

Several strategies are available to reduce honeybee kills. Where possible, avoid pesticide applications during bloom. If spraying becomes necessary, choose insecticides that are least toxic to bees; a selective chemical or a combination of selective chemicals is much less harmful to honeybees than one broad-spectrum insecticide. In addition, the formulation and type of application affect the hazard; wettable powders are often more hazardous than emulsifiable or water-soluble formulations, and sprays of large droplets are more hazardous than those of fine droplets. Spray during the late evening or early morning before honeybees are active to allow the pesticide to dissipate or break down to a level less toxic to bees. Avoid drift and contamination of water sources frequented by bees.

If you have to apply an insecticide that is toxic to honeybees, check with your agricultural commissioner; he will know if beekeepers in your area have requested notification by collect telephone or telegraph. Inform these beekeepers of your scheduled spraying ahead of time so that they can take precautions. Precautions may include confining the bees to the hive by draping the hives with wet burlap or by supplying water in the hives and screening the entrances. Confining bees to the hive can provide sufficient protection for a day or two, although overheating in the hive can occur. If direct spraying of the hives is anticipated, the beekeeper may have to move the hives, which is a labor-intensive operation and often

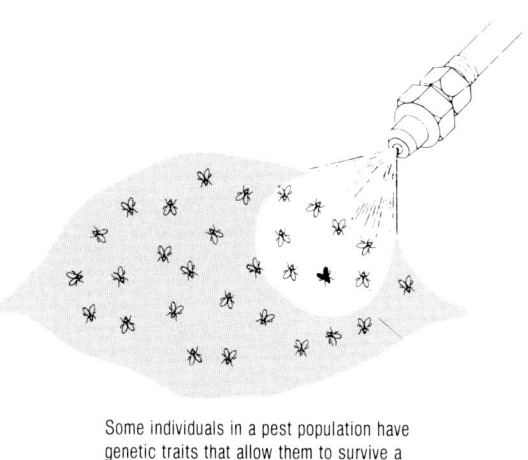
Some individuals in a pest population have genetic traits that allow them to survive a pesticide application.

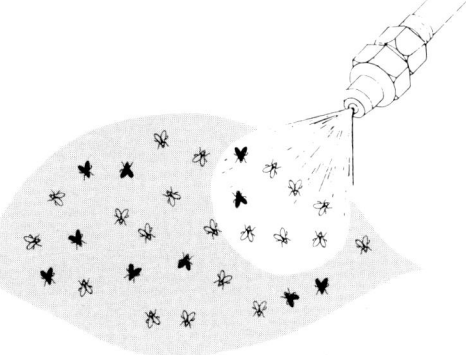
A proportion of the survivors' offspring inherit the resistance traits. At the next spraying these resistant individuals will survive.

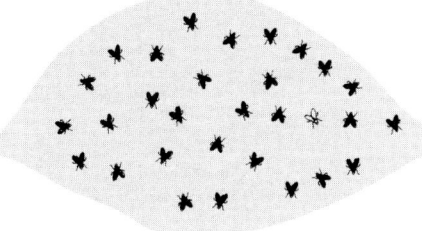
If pesticides are applied frequently, the pest population will soon consist mostly of resistant individuals.

susceptible individual

resistant individual

Figure 27. Pest populations develop resistance to pesticides through genetic selection.

Figure 28. Honeybees forage on pollen and nectar of citrus flowers. If insecticide treatment is necessary during bloom, take precautions to avoid killing honeybees.

presents the problem of finding a new, low-risk location. Communication between the pesticide applicator, agricultural commissioner, and beekeeper can help find an acceptable solution for all parties. For more details on bee protection, see the publications listed above or contact your agricultural commissioner's office.

Phytotoxicity. Many pesticides used in citrus—in particular, spray oils, herbicides, and growth regulators—are phytotoxic if used incorrectly. Spray oils may damage foliage and fruit when applied under dry, hot conditions. Oil sprays should not be applied when soil moisture and humidity are low and temperatures are likely to exceed 35°C (95°F) during the day. The oil film interferes with physiological functions of the plant, such as photosynthesis and transpiration. If the interference persists long enough, soluble solids (sugars) of the juice are reduced, the fruit do not develop good color, and in severe cses leaves may drop and yield may be reduced. The impact of oil treatments varies with the cultivar: the tolerance to oil decreases from lemon to grapefruit, Valencia, navel, tangerine, and lime.

Herbicides, which are designed to kill plants, require particular attention to avoid injury to the citrus trees (see Weeds, p. 133). Growth regulators, such as 2,4 D, can cause leaf curling and chlorosis. For more details and photos of spray injuries, see the disease section.

Hazards to People and Wildlife. Many of the pesticides used in citrus are potentially hazardous to people and wildlife. Applicators are most heavily exposed to pesticides, and they may be at risk of acute poisoning or chronic ailments if adequate safety procedures are not followed. Also at risk are harvesting crews and other orchard workers who may be exposed to toxic residues that may remain in the orchard for many days or even weeks after application. In recent years a number of workers have been injured by toxic residues and breakdown products of the insecticide parathion. Minimize health problems by following label directions, wearing appropriate protective clothing, using closed mixing and loading systems where possible, observing state and local regulations, and confirming the availability of emergency health care. Reentry requirements must be followed carefully. Always post sprayed groves as required by law. To reduce drift, check the wind speed and direction and the calibration of the equipment. Avoid drift into ponds, rivers and other water supplies. Leaflet 2768, *Safe Handling of Agricultural Pesticides*, and 4070, *Pesticide Application and Safety Training Manual*, Agricultural Sciences Publications, discuss health hazards and safety precautions in more detail (see reference list).

Vertebrates

A number of vertebrate species may live within or move into citrus orchards for food or shelter. Pocket gophers and ground squirrels are the two rodents most often responsible for damage. Gophers gnaw the root system or girdle the crown region of young trees; ground squirrels feed on fruit and dig burrows under the trees. Far less important vertebrate pests are meadow mice and rabbits. Meadow mice seldom move into weed-free orchards unless thick mulches and dense tree skirts offer protection. Jackrabbits at times feed on leaves and bark of young citrus trees if other preferred food is not available. Cottontails are not a pest in citrus. Ground squirrels, meadow mice, and jackrabbits also gnaw on plastic irrigation hoses. Roof rats are a concern in some areas because they feed on bark and fruit, and are most common in southern California. Occasionally, wood rats and tree squirrels feed on citrus fruit. Deer may cause some damage to young trees in localized areas near foothills but generally feed on other preferred foods.

The potential for damage by vertebrates varies from orchard to orchard and depends largely on the region, the surrounding terrain, and on the management practices in the orchard. Orchards near rangeland, wooded areas, or other uncultivated areas are more likely to be invaded or reinvaded by certain vertebrates than orchards bordering field crops. Flood irrigation and clean cultivation discourages to some extent infestation by meadow mice, gophers, and ground squirrels.

Predators, diseases, and available food sources all may influence a vertebrate population. Predators, such as coyotes, foxes, badgers, snakes, hawks, and owls, feed on many rodent and rabbit species. Such predators, however, cannot be relied on to prevent rodents or rabbits from becoming agricultural pests at times.

Management Methods

A few preventive measures can make the orchard or the surrounding areas less favorable to invasion by vertebrate pests and survival. For example, clearing weeds and thick mulches around tree trunks will discourage infestation by meadow mice. If the preventive approach proves unsatisfactory or unfeasible, other management methods,

Figure 29. Mounds without an opening are typical of pocket gopher activity.

such as trapping, baiting, fumigating, or shooting, depending on the pest, may have to be used.

Traps can be effective for some rodent species, such as pocket gophers and ground squirrels. Poison bait is available for rodent pests and jackrabbits as single-dose or multiple-dose bait. Single-dose baits kill the animals after one feeding. Several common, single-dose poisons produce bait shyness with some pests if the animal ingests only a sublethal dose. Prebaiting with untreated bait may assure a better initial control. Most multiple-dose baits are anticoagulants, which act more slowly than single-dose baits and in general require several feedings over a period of days. These baits kill animals by causing internal bleeding; they are not likely to cause bait shyness. Multiple-dose baits are usually considered less hazardous to humans and some nontarget species because they produce delayed symptoms, giving time to administer an antidote. Contact your county agricultural commissioner for information on available baits. Follow label directions for rates, application methods, and precautions.

Fumigation is generally limited to ground squirrel control; it is not very effective against pocket gophers or meadow mice in citrus. Shooting may be feasible for a small population of ground squirrels or jackrabbits. Table 8 summarizes the legal status and control restrictions for common vertebrate pests in citrus.

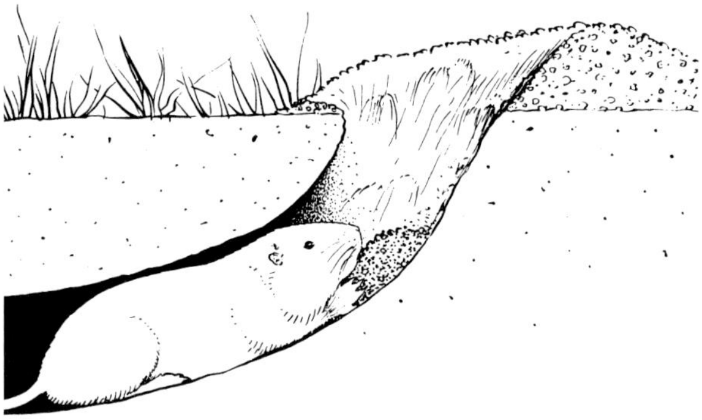

Table 8. Legal Status and Control Restrictions of Common Vertebrate Pests in Citrus

Species	Status and Restrictions
Pocket gophers, Ground squirrels, Meadow mice, Roof rats, Wood rats	Classified as nongame mammals; can be killed in any manner at any time if injuring crops.
Jackrabbits	Classified as game mammals by California Fish and Game Code; hunting season is all year with no bag limit. When injuring crops, jackrabbits may be killed in any manner, but traps must comply with Fish and Game regulations.
Tree squirrels	Classified as game mammals by California Fish and Game Code; a permit from the regional office or local game warden required for control of the eastern gray squirrel; poisoning of this species is illegal. The eastern fox squirrel may be killed in any manner if causing damage.
Deer	Classified as game mammals; a depredation permit is required for shooting. Poisoning of deer is illegal.

Figure 30. Pocket gophers dig their tunnels by pushing soil out through lateral tunnels. They plug the openings with soil when completing the tunnel.

Because each mammal pest is managed differently, it is important to identify correctly the species that is causing the damage. Watch for the animals or indications of their presence, such as burrows, soil mounds, feces, gnawing marks on the limbs or trunk, or citrus peel and partly eaten fruit. Ask your agricultural commissioner or farm advisor for help if you are not certain about identification.

The choice and timing of a control action is often critical. For instance, you can control ground squirrels most effectively after they have emerged from hibernation and before they have reproduced. At this time in early spring, fumigation is the best choice because the soil is moist and retains the toxic gas; grain bait is not readily accepted because the squirrels feed mainly on fresh vegetation at that time of year.

Because of the complexity of vertebrate pest problems, no numerical guidelines are available that specify what infestation levels and combination of environmental conditions justify control actions. Instead, to make management decisions for a vertebrate pest, the biological information presented in this chapter and the history of vertebrate pests in your orchard should be weighed against the cost and effectiveness of available control methods. After you have achieved control, establish a routine monitoring program to detect reinvasions and to make future control decisions easier and more efficient.

Pocket Gophers
Thomomys spp.

Pocket gophers are important vertebrate pests in citrus orchards. They cause damage by feeding on young tree roots and gnawing the bark of the trunks just below the soil line, often girdling young trees. Their burrows may divert irrigation water and contribute to soil erosion.

Pocket gophers spend most of their time underground and are rarely seen, but characteristic crescent-shaped mounds of soil indicate their presence (Figure 29). The gophers dig tunnels 10 to 30 cm (4 to 12 inches) beneath the surface, and push the soil out lateral exits and plug them (Figure 30). In citrus areas, gophers breed in early spring, and the females may bear a litter of five to six young. If conditions are favorable, they may produce several litters a year.

Management Guidelines

Look for evidence of gophers in late fall, winter, or spring. As the ground becomes moist from rain, the gophers increase their digging activity, and you can easily detect fresh mounds. In late spring and summer, you may find gopher mounds in new parts of the orchard, as the young set up burrow systems of their own. About once a month, check under the tree skirts, especially in border tree rows where gophers may move in from adjacent fields or orchards.

Control gophers as soon as you detect their presence; a single gopher can severely damage a tree. Trapping or baiting by hand is most effective. Machine baiting is generally not used in citrus because it disturbs the top soil, destroying feeder roots in established orchards or interfering with irrigation systems. Fumigants do not work well because the burrow system is extensive, and gophers can quickly seal their tunnel when they detect the poisonous gas.

Traps are useful when a limited number of gophers is present, and they are most effective if inserted into the main tunnel between two fresh mounds. Locate the main tunnel, which usually extends from the lower side of the mound, with a steel rod 6 mm (¼ inch) wide or a specially designed metal probe. Place traps as shown in Figure 31. For the most effective control, inspect the traps twice daily.

You can also control pocket gophers with single-dose poison bait. Check with your agricultural commissioner for availability of baits. Locate the main burrow, enlarge

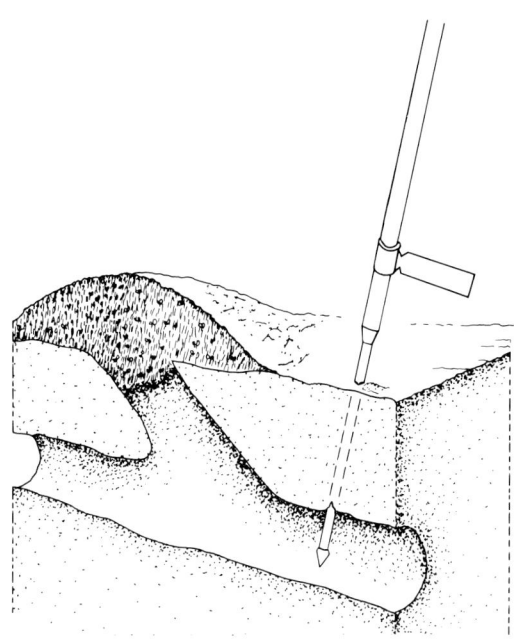

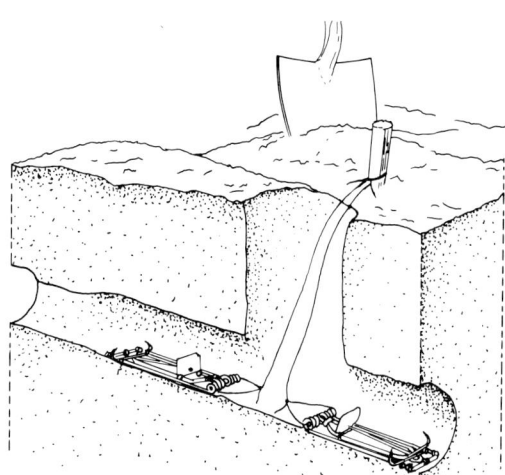

Figure 31. Control of pocket gophers using traps. Locate the main burrow with a metal probe (top); place two traps in opposite directions, fasten them to a stake, and cover the hole (bottom).

Figure 32. The subspecies of the California ground squirrel are similar in appearance, except that *Spermophilus beecheyi douglasii* has a dark patch or stripe on its back (top). This marking is absent from *S. beecheyi beecheyi* (bottom).

the probing hole with the metal rod, and drop the bait into the tunnel, following label instructions. Cover the hole but prevent soil from dropping onto the bait. Repeat once or twice along the runway. Handle poison bait with extreme care to avoid spilling bait on the ground surface, where it may cause accidental poisoning of nontarget animals. Keep bait out of reach of pets and children.

California Ground Squirrel
Spermophilus beecheyi

Three subspecies of the California ground squirrel are found in the citrus-growing areas of the state. *S. beecheyi beecheyi* occurs along the coast from San Francisco southward, *S. beecheyi fisheri* inhabits the San Joaquin Valley, and *S. beecheyi douglasii* lives in the Sacramento Valley. Management guidelines are the same for these subspecies, except when using single-dose baits.

Ground squirrels may feed on ripe fruit on the tree skirts, consuming the entire fruit or just eating the seeds. Generally, however, they cause more damage by digging burrows under the trees and gnawing polyethylene irrigation hoses. In some areas, ground squirrels present a potential health problem because they carry fleas, which may transmit infectious diseases.

The California ground squirrel weighs 0.5 to 0.7 kg (1 to 1.5 lbs) as an adult. It is about the same size as a tree squirrel but has a much less bushy tail and a mottled fur with flecks. *S. beecheyi douglasii* has a dark patch on its back (Figure 32); *S. beecheyi fisheri* looks much like *S. beecheyi beecheyi*. Ground squirrels often bask in the morning or late afternoon sun. They are social animals, living in groups of up to 20 individuals.

Although ground squirrels may not be seen, their burrows are an easily recognizable indication of their presence. Each burrow system has several open entrances, usually with scattered soil in front (Figure 33). (Burrows of pocket gophers are smaller in diameter and have plugged entrances with crescent-shaped mounds.) Individual ground squirrel burrows may be 1.5 to 9 m (5 to 30 feet) long, 75 cm to 1.2 m (2.5 to 4 feet) below the surface, and about 10 cm (4 inches) wide. Ground squirrels usually dig their burrows along ditches, fence rows, and on uncultivated land. They may also establish burrows beneath trees in the orchard.

Ground squirrels emerge from hibernation about late January in the southern part of the Central Valley and about a month later in the northern part. The females have one litter of six to eight young in the spring. About six weeks after birth, the young emerge from their burrows to feed above ground. Ground squirrels prefer to feed on green vegetation in the spring, and on seeds, grains, fruits, and nuts during the summer and fall.

The adults often go into a temporary state of inactivity (aestivation) for part of the hot summer and into hibernation in the winter. The young may not aestivate or hibernate during the first year. Because of these periods of inactivity, control actions have to be timed accordingly. (Figure 34 shows the major activity periods and food sources of ground squirrels.)

Management Guidelines

Whether and when you need to take management action depends largely on the local conditions. If your orchard is surrounded by infested, uncultivated land, reinvasion will be a continuous problem. In cultivated areas where highly desirable habitat is limited, removing old stumps, brush piles, and debris can be helpful in restricting populations.

To monitor ground squirrels, check the perimeter of the orchard about once a month for animals or their burrows. Check during the early morning or late afternoon when squirrels are most active. Record the approximate number of squirrels and the location and number of burrows.

If your monitoring records indicate that the squirrel population is moving into the orchard, control them with traps, fumigants, or toxic bait. Choose the method carefully according to the time of year (Figure 34), area, and number of squirrels, as well as the cost, labor, and potential hazard of the treatment.

You can control ground squirrels most effectively in late winter or early spring after they have emerged from hibernation and before they have reproduced. In timing your control actions, keep in mind that males appear first; all animals will have emerged from hibernation when the sex ratio is 1:1. At this time you may either use a fumigant or traps that do not require baiting. Baited traps or toxic bait provide good control later in the spring or in the fall when squirrels collect and eat seeds and fruit. If you are in doubt about correct timing for bait acceptance, consult your agricultural commissioner. Attempts at controlling ground squirrels during the summer are usually not successful because many of them are aestivating.

Fumigants. Fumigants work best in the early spring when moist soil helps retain a high toxic gas level in the burrows. Several types of fumigants are available. Nonrestricted fumigants, such as gas cartridges (woodchuck or smoke bombs) are available from agricultural commissioners or private companies. When ignited, these cartridges release smoke and toxic gases, in particular, carbon monoxide. Following label instructions, place one or more cartridges in the active burrow opening and plug all other openings with soil. Insert and ignite the fuse, push the cartridge deep into the burrow, and quickly seal the entrance. Be careful not to inhale the smoke. Watch the area for escaping smoke and plug any leaks. Check the burrows after about three days. Treat again where opened burrows indicate that squirrels have dug out. Do not use cartridges

Figure 33. Entrances to ground squirrel burrows are open.

where fire may be a hazard, such as near dry grass or beneath a building.

Restricted fumigants, such as aluminum phosphide, carbon bisulfide, and methyl bromide, require a permit from the agricultural commissioner. These materials, if misused, are hazardous and should be applied only by experienced persons. Carbon bisulfide and methyl bromide are toxic to plants and can damage or kill citrus if applied to burrows beneath trees. Apply these materials only to orchard margins away from trees, and strictly follow all label instructions. Smoke bombs and aluminum phosphide do not harm trees.

Traps. Trapping ground squirrels works well in small areas or for a small number of squirrels. Be sure to place traps in such a way that pets and wildlife are unlikely to be caught.

One effective trap is a box-type pocket gopher trap, which can be modified to catch ground squirrels (Figure 35). Place a bait of citrus, melon rind, grains, or nuts between the two traps under the wire mesh to lure the squirrels through the trapping mechanism. Bait but do not set

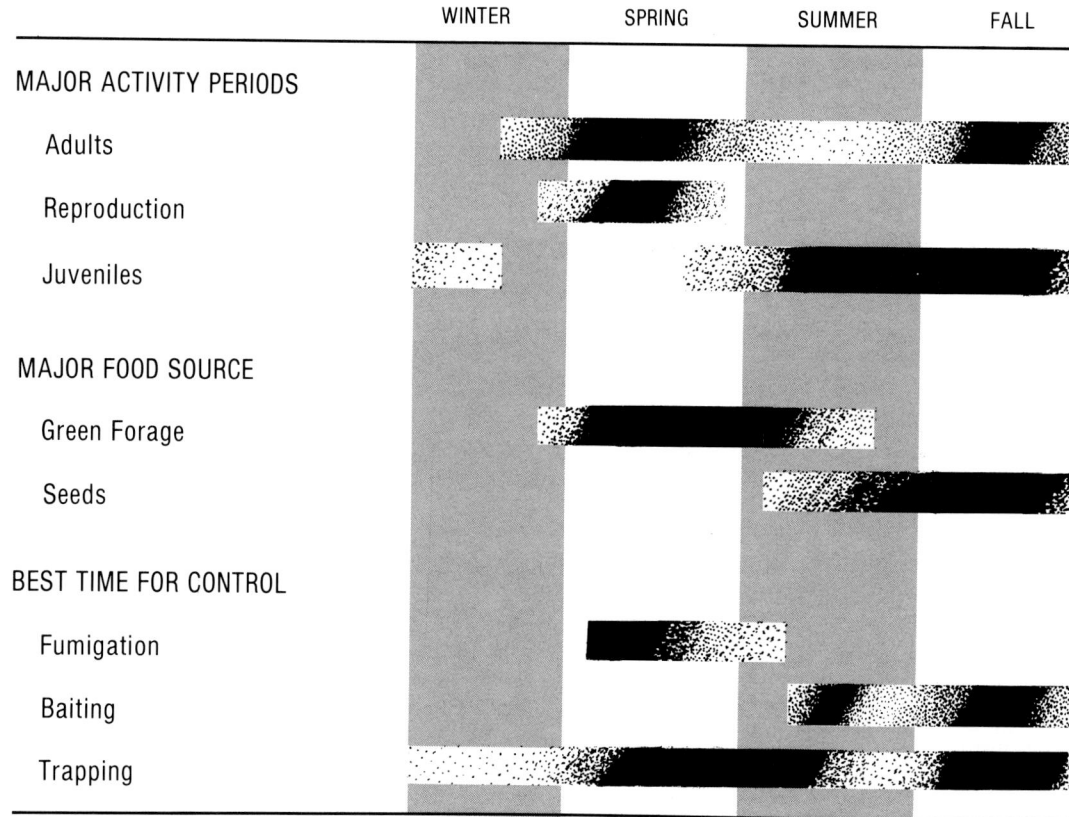

Figure 34. To choose the best time and type of control for the California ground squirrel, consider its activity periods and food sources during the year.

the trap for a few days until the squirrels have become accustomed to feeding at the station.

Another trap, the Conibear®, does not require baiting if set over the burrow entrance. Live traps work well but are expensive, and the live animals have to be destroyed after they are caught.

Toxic Bait. Poison baits, either single dose or multiple dose, are the control method most commonly used against ground squirrels. Check with your agricultural commissioner for the most effective bait for the subspecies occurring in your area. Depending on label instructions, single-dose bait is generally distributed by hand over 0.2 to 0.3 square m (2 to 3 square feet) near burrow openings. Do not pile bait; spreading it thinly takes advantage of the squirrels' natural foraging habits. Depending on the toxicant and the situation, remove and dispose of carcasses after a treatment to prevent poisoning dogs or other scavengers that may feed on them.

Multiple-dose baits are often registered for use in bait stations. You can make bait stations from metal or plastic pipe, discarded car tires, or wood (Figure 36). They should have entrance holes large enough to admit squirrels but small enough to prevent larger animals from entering. Place the station near runways or burrows and secure them so that they cannot be tipped over easily. When you first set them out, inspect the stations daily and add bait as needed. Maintain the bait until feeding ceases. This process may take one to four weeks because many squirrels may not start feeding immediately, and individual susceptibility to anticoagulants varies. Remove the unused bait from the stations once control is completed.

Some anticoagulants are registered also for spot baiting or broadcasting. Lightly scatter bait near burrow openings or runways and reapply every other day for at least three applications or as indicated on the label.

Meadow Mice
Microtus spp.

Meadow mice, also called voles or field mice, inhabit roadsides, meadows, canal banks, fencerows, and many field crops. They are rarely a problem in weed-free citrus orchards, although occasionally some damage may occur at the edges bordering field crops or other preferred habitats. Young orchards with some weed cover are more likely to have a mouse population, especially if established in former uncultivated land or rangeland. Meadow mice feed on young or mature trees, gnawing through the tree bark to the tender cambium layer (Figure 37). Young trees may be completely girdled from close to the soil line to as far as the mice can reach.

Full-grown meadow mice are larger than house mice but smaller than rats (Figure 38). Well-established populations can usually be recognized by the network of small

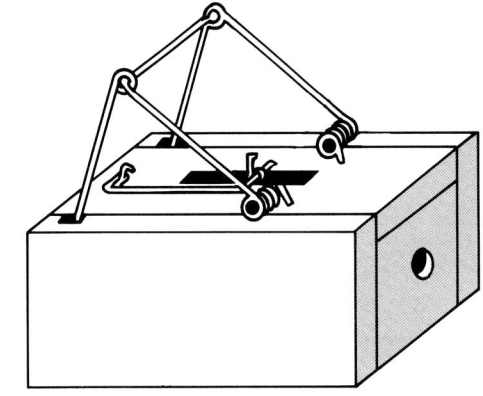
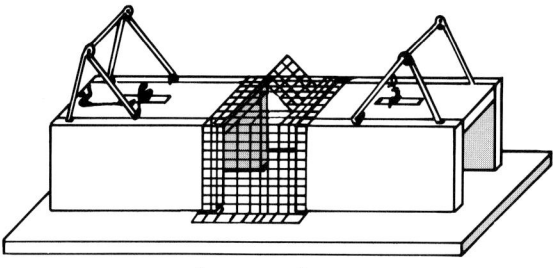

Figure 35. Single, box-type pocket gopher traps are effective for catching ground squirrels and tree squirrels. To use them in pairs, remove the back of the traps (top), connect two traps with wire mesh, and attach them to a board (bottom).

Figure 36. You can make a bait station for ground squirrels out of wood, discarded tires, or plastic pipe, such as this one.

Figure 37. Meadow mice gnaw through the bark of young or mature trees, partly or completely girdling the trunk from the soil line upward as far as they can reach.

Figure 38. Meadow mice have a blunt nose, small eyes and ears, and a short tail. They are entirely gray or brown and may be up to 15 cm (6 inches) long, not including the tail.

runways through grass or other cover and the openings of numerous shallow burrows. Meadow mice are active all year, day and night. Females may bear four to eight litters each year with a peak reproduction period in the spring. Uncontrolled populations fluctuate dramatically, reaching a peak every four to seven years, then declining sharply.

Management Guidelines

Certain cultural measures help prevent damage from meadow mice. Where orchards are kept free of weeds, meadow mice generally do not move in. When maintaining a ground cover, keep weeds at least 90 cm (3 feet) from tree trunks. Do not let mulches build up under trees. Clear weeds from surrounding fencerows or ditch banks. Tree wrappers used for sunburn and frost protection offer some protection against meadow mice (Figure 39), although wrappers may at times offer shelter for the deer mouse or house mouse. In the fall or winter, check the orchard and surrounding land for signs of mouse activity. If you find mice, treat with bait before spring breeding season.

Hand baiting gives the most effective control in citrus. Place bait in active runways, burrow entrances or at several spots around the trunks of trees. Where large, young orchards are infested, machine broadcasting may be more economical. Baiting from the air is not suitable for mouse control in mature orchards because of the dense canopies.

Both single- and multiple-dose baits are effective for mouse control. After a treatment, monitor the orchard carefully for surviving or migrating meadow mice that may reestablish themselves. A follow-up treatment may be needed if the population begins to build again.

If young trees are threatened by repeated mouse invasions, wire guards may be feasible. Place a mesh hardware cloth, 0.6 to 1.3 cm (¼ to ½ inch) mesh size and 60 cm (24 inches) wide, around the trunks, leaving enough space for several years' growth. The guards must extend at least 15 cm (6 inches) below the soil line. Where jackrabbits are a problem, make the guards 90 cm (36 inches) high for added protection.

Black-Tailed Jackrabbit
Lepus californicus

The black-tailed jackrabbit is a common agricultural pest and may feed on the bark of young trees. As a rule, jackrabbits cause little damage to mature citrus. But they often seek the shelter of the orchard to rest during the day, and they move out to forage on field crops during the night.

In recently established orchards, tree wraps against sunburn and frost also offer some protection against jackrabbits. Where jackrabbits and meadow mice are a constant threat to young trees, wire guards may be feasible (see section on meadow mice). Jackrabbits can be killed in any manner when they cause agricultural damage (Table 8). Jackrabbits will not readily enter live traps, and shooting may be the best solution where only a few animals need to be removed. Baiting is rarely successful. Check with your agricultural commissioner for the most effective and economical protection of your trees.

Other Vertebrates In Citrus

The root rat (*Rattus rattus*) is an introduced species that is mainly found near dwellings but may also occur in citrus orchards (Figure 40). It has been a particular problem in certain areas of southern California. Roof rats build nests in citrus trees and eat the pulp out of mature oranges or lemons. If the rats are numerous, they may also chew the bark of scaffold limbs. Roof rats breed several times a year, producing litters that average about six young.

Roof rats are most visible in border trees, where they often first establish their nests. To control roof rats, single-dose and multiple-dose baits are available. Anticoagulant bait can be offered in bait boxes or ready-made paraffin blocks secured to limbs 2 m (6 feet) or higher. Trapping can be effective for a few rats. Fasten rat snap traps to limbs and bait with citrus, raisins, prunes or nut meat. Do not set traps until bait is readily accepted.

The wood rat (*Neotoma spp.*), also called pack rat, is a native rat resembling the roof rat, but its fur is softer and the tail is hairier. Wood rats normally live in wooded or brushy areas. They may move to adjacent citrus trees to feed on fruit or bark and cut twigs for their large nests. Wood rats build their nests on the ground or in trees in wooded areas, but in citrus orchards, nests are always built in trees. When control is necessary, the same methods as used for the roof rat will be effective.

The cotton rat (*Sigmodan hispidus*) has a limited distribution and may be found on grassy ditchbanks of interior southern California. It rarely causes economic damage to citrus. Appearance, habitat, damage, and control measures are similar to that of the meadow mouse.

Two introduced species of tree squirrels occur in some citrus-growing districts. The eastern fox squirrel (*Sciurus niger*) is more common than the eastern gray squirrel (*Sciurus carolinensis*). The fox squirrel is well established in many city parks, residential areas, and adjacent agricultural lands. The eastern gray squirrel is more localized and rarely causes damage to citrus.

Figure 39. Tree wrappers offer some protection against meadow mice and, if high enough, against jack rabbits.

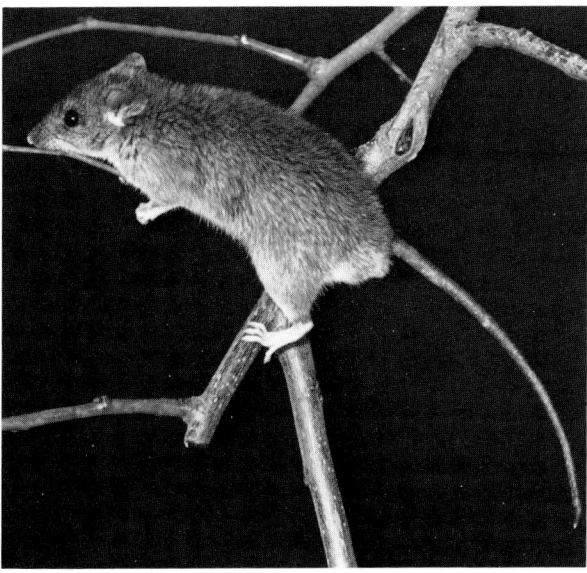

Figure 40. The roof rat has large eyes and ears and a pointed nose. Its tail is longer than its body and head combined.

Figure 41. The eastern fox squirrel has a bushy tail and reddish brown fur.

Tree squirrels have a bushier tail than ground squirrels, and their fur is not flecked or mottled (Figure 41). The eastern fox squirrel is reddish brown; the eastern gray squirrel has a grayish color similar to the native western gray squirrel (*Sciurus griseus*). Tree squirrels live in woody areas along creeks and river banks, from where they move into adjacent orchards for food. The eastern fox squirrel feeds on ripe citrus fruit, nuts, mushrooms, and on bird eggs and insects. Tree squirrels are classified as game mammals, and their control is regulated by the California Fish and Game Code (Table 8). When legal, trapping is usually sufficient to control tree squirrels. The modified pocket gopher box trap (Figure 35) can be used for tree squirrels. Fasten one or two box traps on a horizontal limb in a tree where feeding damage occurs and bait it with pieces of nut meat. Bait but do not set traps for several days to allow the animals to get used to the traps.

Deer occasionally damage newly planted citrus that is located near natural habitats, such as woods and streamside thickets. They feed at night on young tree foliage and rub their antlers on limbs in the spring. Citrus trees, however, are less desirable food for deer than certain other fruit and nut trees. Foliar repellants are available and offer some protection. If deer control is necessary, contact the Fish and Game office or your local game warden—they sometimes issue depredation permits to shoot deer that damage crops. Fencing, although more expensive, offers a more permanent solution.

Insects, Mites, and Snails

Climatic and geographic conditions have a great influence on the invertebrate pests and natural enemies present in the various citrus-growing areas and on the options available for pest management. The areas contrasting most in climate are the San Joaquin Valley, with its hot summers, cold winters, and seasonal fluctuations in humidity, and the coastal-intermediate areas of southern California, with their mild temperatures and higher humidity. The interior districts of southern California are warmer and dryer than the coastal-intermediate areas. Fairly constant, low humidities and extremes in daily temperatures characterize the desert valleys of southern California.

The San Joaquin Valley

In the San Joaquin Valley, citrus thrips and California red scale are the key pests; natural enemies do not effectively contain these pests in this region. Citrus thrips scar rind tissue superficially but do not affect the internal quality of the fruit. Blemished fruit, however, bring lower prices, and treatments are applied when thrips populations reach a threshold level during the critical injury period, from petal fall until young fruit are about 4 cm (1.6 inches) in diameter. The California red scale is also a cosmetic pest at low population densities, but populations have the potential to increase rapidly and damage the tree. Development of monitoring techniques and threshold levels have reduced the number of treatments necessary to control this pest.

Other important pests are the orangeworms, especially the citrus cutworm and the fruittree leafroller. The larvae of these moths feed on young foliage, blossoms, and fruit. Monitoring and management guidelines have been developed to control these pests without destroying honeybees and natural enemies. The citrus red mite is a potentially damaging pest every year. Predaceous mites, a viral disease, and hot, dry weather may keep the citrus red mite below damaging levels; careful monitoring, however, is necessary to determine the need for treatment and avoid losses.

Natural control factors completely or partially control certain other pests in the San Joaquin Valley. The cottonycushion scale and the brown soft scale are under

excellent biological control, if undisturbed by insecticides or adverse weather. The citricola scale lacks adequate natural enemies in the San Joaquin Valley but is usually well controlled by sprays applied for California red scale.

Insect and mite control in the San Joaquin Valley rely mainly on the use of synthetic organic chemicals. The synthetic chemical program limits the effectiveness of natural enemies of other pests, such as mites and scales, and further treatments are often needed. Depending on the monitoring results, a seasonal treatment program may include a prebloom or postbloom application against orangeworms or citrus red mite, or both, a petal-fall treatment for citrus thrips, and a June treatment for the California red scale. In some years, no treatment for the citrus red mite or the California red scale may be necessary.

The Coastal-Intermediate and Interior Districts

In the coastal-intermediate growing area of southern California, mild climatic conditions allow effective biological control of many citrus pests. Biological control is somewhat less effective yet still important in the southern interior district. Two parasitic wasps control the California red scale in many situations. A number of pest species, previously of major concern, are now generally under good biological control. These include the cottonycushion scale, black scale, brown soft scale, purple scale, mealybugs, and most whitefly species. A recent invader, the bayberry whitefly, presents a threat to the existing biological control program, but efforts to establish natural enemies to control it are very promising.

Several other pests are of concern because they have insufficient or no biological control. The citrus thrips, which is a particular problem on coastal lemon, is currently not under good biological control, although *Euseius hibisci*, a predator of the citrus red mite, and other general predators have been observed to feed on thrips larvae. *E. hibisci*, general predators, and a viral disease usually suppress the citrus red mite in the coastal-intermediate district unless upset by chemical treatments. In southern interior districts, however, the citrus red mite often requires treatment.

Three other mite species not found in the San Joaquin Valley are frequently pests in the coastal-intermediate district. The citrus rust mite (silver mite) causes a russeting of oranges and a silvering of lemons on the coast, and the citrus bud mite and broad mite occur only on lemons. The broad mite has become a problem in recent years in coastal orchards.

A management program in coastal-intermediate and interior distrists of southern California can rely on many natural enemies and encourage and supplement their actions. With few exceptions, only supplementary treatments with oils or selective insecticides are needed. Because of the area's mild climate, oil has long been used successfully in the control of the California red scale, other scales, and mites. In recent years, low-volume oil sprays have begun to replace dilute oil sprays. Low-volume applications use less oil and water, thus reducing substantially cost and phytotoxicity while achieving effective control. Low-volume oil sprays are also much less destructive to adult parasites than are dilute sprays. When oil provides insufficient control, choose selective chemicals for pests such as citrus thrips, broad mite, and rust mite.

Decisions to use insecticides and miticides should always be based on the results of monitoring. In some groves, no applications may be required. An application against citrus thrips may be needed at petal fall on oranges and during June or July on lemons; the botanical material sabadilla can provide sufficient control if an infestation is not too heavy. During the late summer or fall, rust mite or broad mite populations may require the application of a miticide. In late summer or early fall, a low-volume oil spray is often applied against the citrus red mite or the California red scale. In certain areas, *Aphytis*, the parasite of the California red scale, and *Cryptolaemus*, the mealybug predator, are released to supplement naturally occurring biological control.

The Desert Valleys

The desert growing regions have the fewest insect and mite problems of all citrus districts. The citrus thrips is the most important desert pest requiring treatment with synthetic organic chemicals. The California red scale is widespread in parts of the Imperial Valley and has become established in a few pockets in the Coachella Valley, where an eradication and quarantine program keeps the scales at extremely low levels. The citrus red mite has become adapted to the desert climate but is rarely a problem. The Yuma spider mite and the flat mite are frequently seen in orchards but populations rarely build up except where natural enemies have been destroyed by treatments for other pests.

Monitoring Insects and Mites

A good monitoring program for insects and mites is essential for implementing integrated pest management. Although they are at various stages of development, monitoring programs are available for orangeworms, California red scale, citrus thrips, and citrus red mite. Use your monitoring results together with orchard history and weather data to determine the need for management actions. It is important to correctly identify the pest species, their associated damage, and their natural enemies because even closely related species may require different control measures. The photos and descriptions in this section will assist you in identifying the most common insects

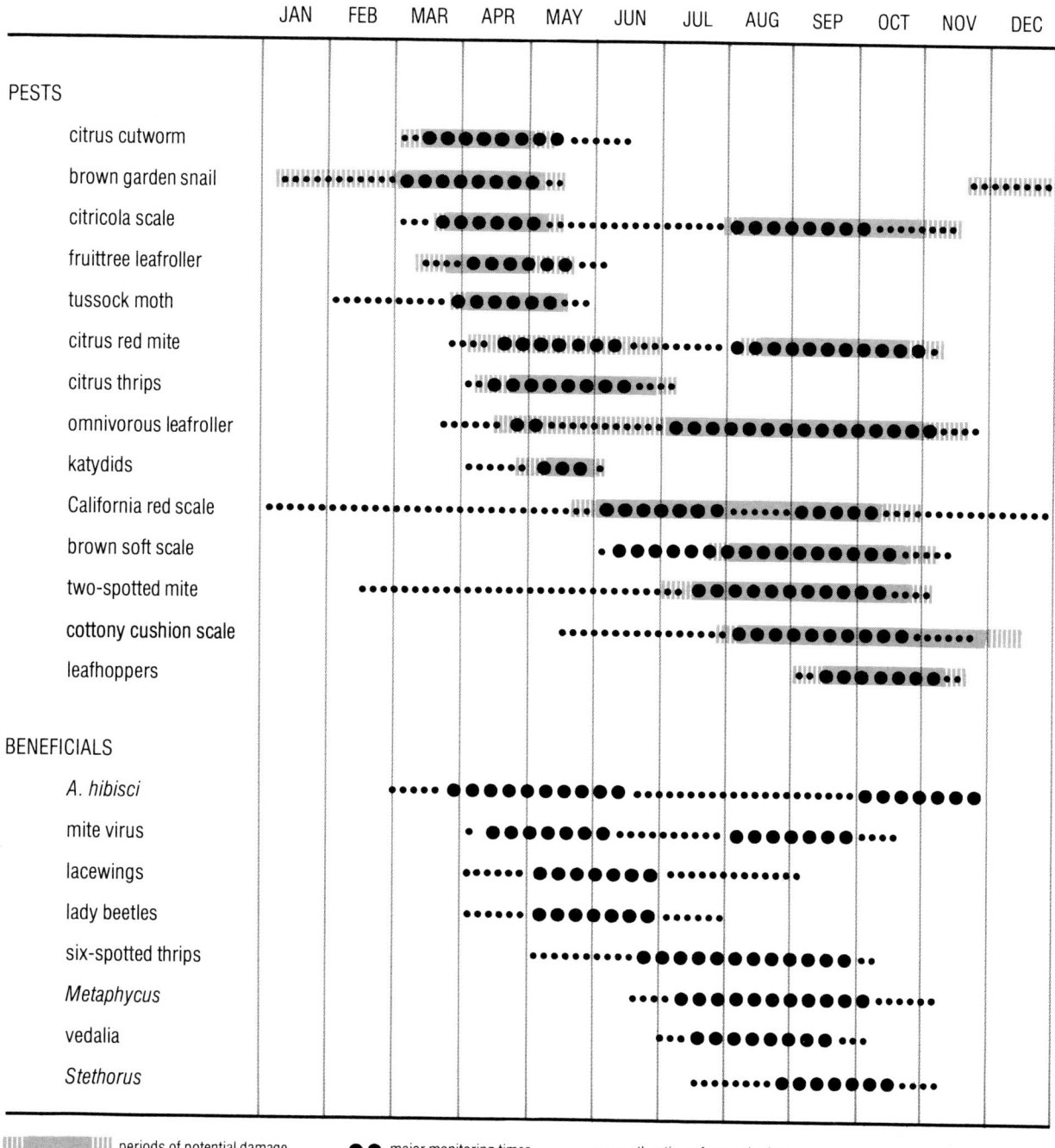

Figure 42. Monitoring and damage periods for insects and mites on navel oranges in the San Joaquin Valley. The brown garden snail has been included because it is a frequent pest requiring careful monitoring. Note that the preferred monitoring times often do not coincide with the times major damage occurs.

VALENCIAS, COASTAL-INTERMEDIATE

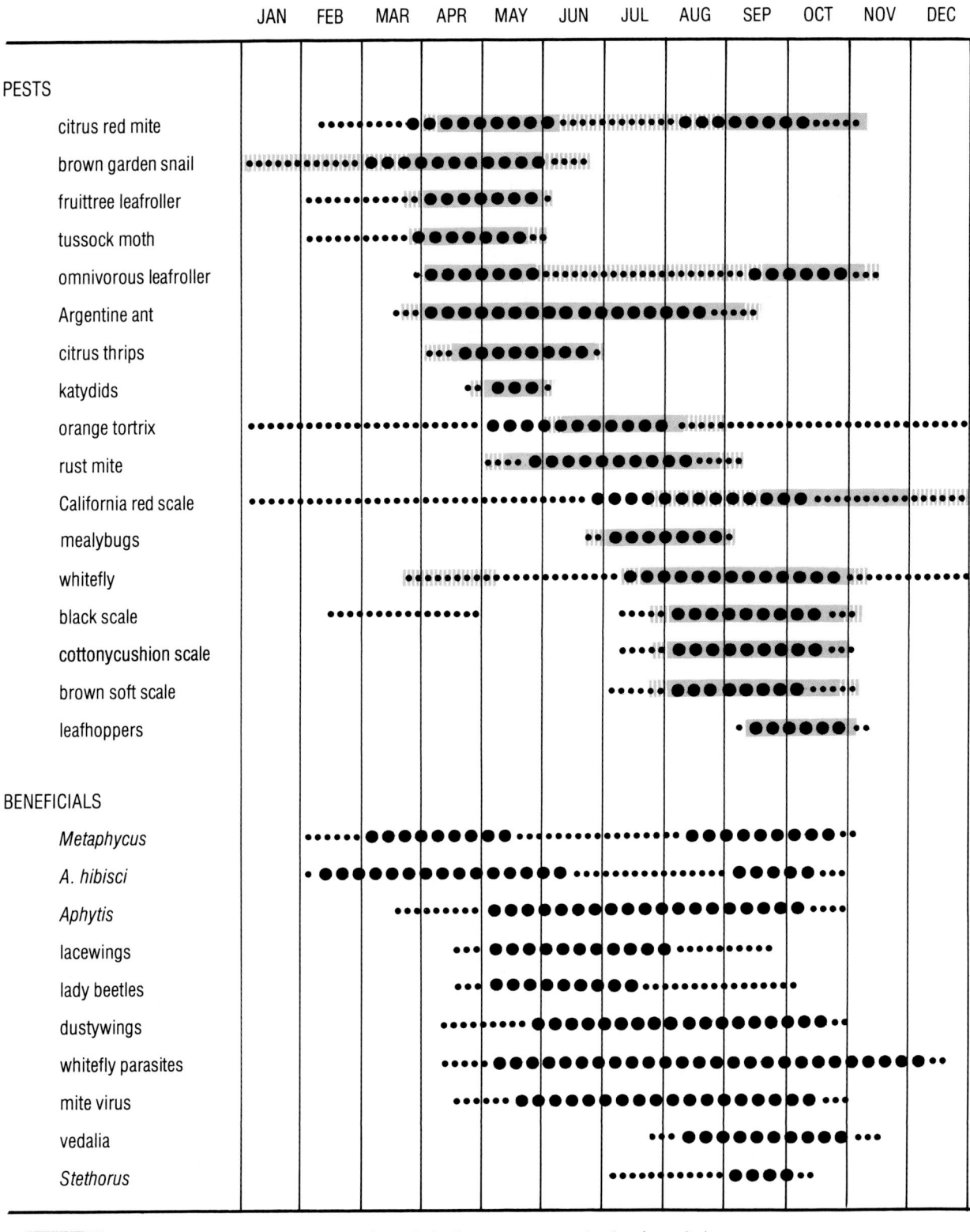

Figure 43. Monitoring and damage periods for insects and mites on Valencia oranges in the coastal-intermediate district. The brown garden snail has been included because it is a frequent pest requiring careful monitoring. Note that the preferred monitoring times often do not coincide with the times major damage occurs.

LEMONS, COASTAL-INTERMEDIATE

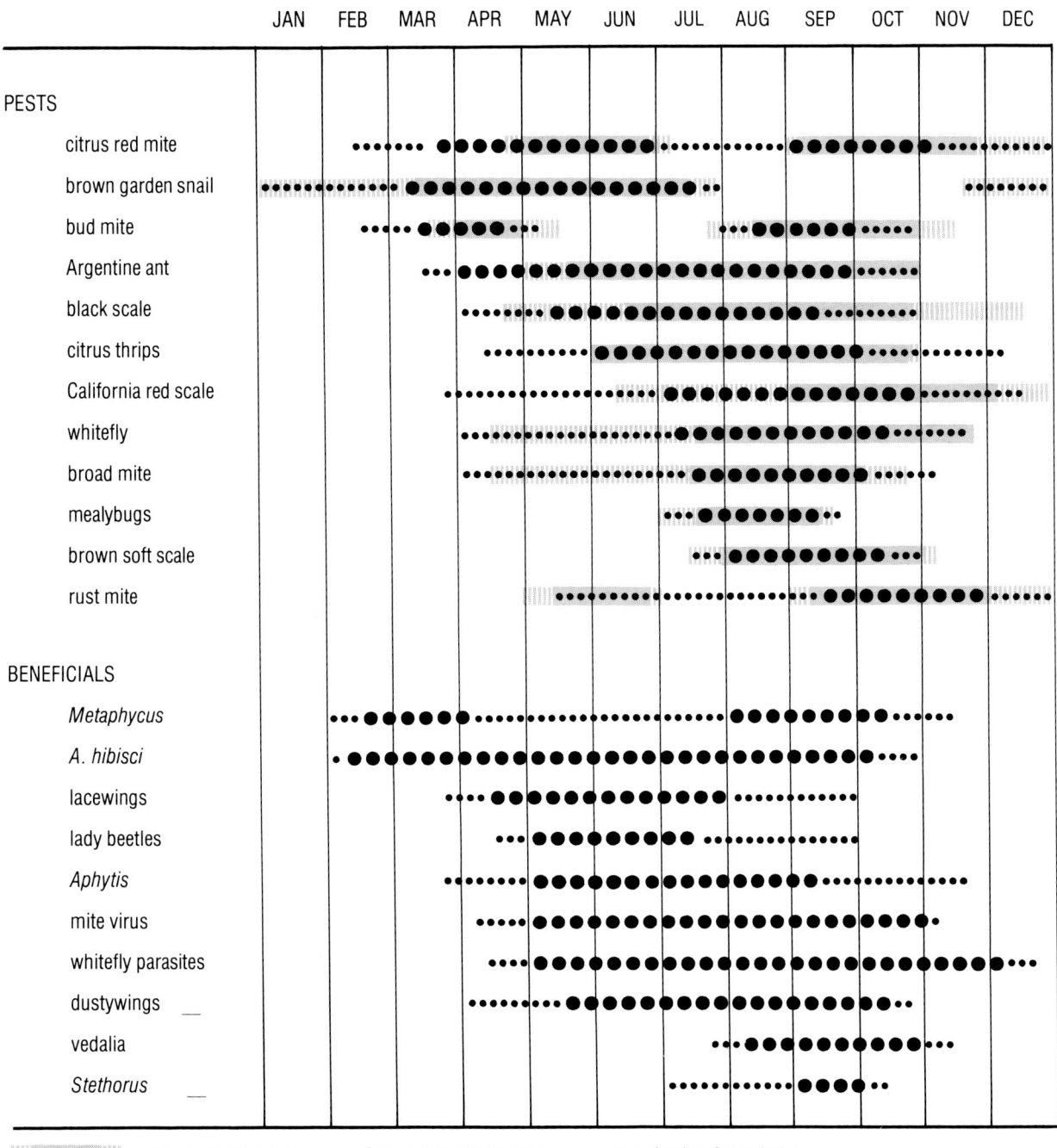

Figure 44. Monitoring and damage periods for insects and mites on lemons in the coastal-intermediate district. The brown garden snail has been included because it is a frequent pest requiring careful monitoring. Note that the preferred monitoring times often do not coincide with the times major damage occurs.

and mites found in citrus orchards. For additional help, consult your farm advisor or PCA.

You need to carry out precise sampling programs for individual species only at certain times of the year. Figures 42 through 44 show the best monitoring times for major invertebrate pests in the San Joaquin Valley, the coastal-intermediate districts, and the interior valleys. The brown garden snail has been included because it is a frequent pest requiring careful monitoring. Note that the preferred monitoring times often do not coincide with the times major damage occurs.

A sampling scheme should include all sections of the orchard. Choose your decision-making unit or "block" according to the size, geography, and microclimate of the orchard. Conditions within the block should generally be uniform. On the coast, the block size may vary from 4 to 6 hectares (10 to 15 acres); in more uniform growing areas, a block may be as large as 20 hectares (50 acres). Select several sampling sites in each block at random (see *Monitoring Methods* for details). Monitor "hot spots" separately, such as the edges of an orchard, where dust or insecticide drift may lead to a buildup of some pests. Record your counts of pests and beneficial species; you can use field record sheets while you are monitoring (examples are given in the individual pest sections) and a summary sheet, such as the one given in Figure 45, to follow the seasonal status of all pests in one orchard or block. Use your own rating system or a scale of 1 to 7, in which 1 is the lightest and 7 the heaviest infestation:

1—**Trace.** Infestation is extremely difficult to find, there are only a few isolated individuals in the orchard, and all trees are below the economic level.

2—**Trace to light.** Infestation generally is still very difficult to find. A few trees within the orchard have an infestation that is relatively easy to find.

3—**Light.** Infestation is usually easy to find, but there are only a few individuals per tree. It is usually not economical to treat at this level; in no case has damage already occurred, and withholding treatment for heavier populations is recommended.

ORCHARD _____ **BLOCK** _____ **CULTIVAR** _____

COMMENTS (pesticide use, etc.) _____

_____

Date	CITRUS RED MITE			CITRUS THRIPS			CALIFORNIA RED SCALE			OTHER PESTS		
	females	predator	virus	% scarred fruit	% infested fruit	predator	% infested fruit	trap counts	parasite			

Figure 45. Example of a summary record sheet for monitoring a citrus orchard. You can copy this form or make your own to keep seasonal records of the citrus red mite, citrus thrips, California red scale, other pests, and their natural enemies. To take notes while you are making counts, use worksheets such as the ones given in the individual pest sections.

4—Light to medium. Infestation is easily seen, with some parts of the orchard approaching the medium level. Infestation is normally increasing with only minor damage occurring. A treatment is now considered.

5—Medium. This infestation is at the economic threshold. Some damage is occurring. At this level treatment should be made as soon as possible.

6—Medium to heavy. At this infestation level, pest species are easy to see. Many individuals occur in a given area. General economic damage is visible. Treatment applied now is usually late.

7—Heavy. Extreme infestation is obvious, with numerous individuals in a given area. Great economic injury has already occurred.

When recording your pest infestation, also note weather conditions, previous management actions, and, where appropriate, the level of predator or parasite activity. These data are essential for making sound management decisions, and they provide valuable historical data for future seasons.

Monitoring Methods

Monitoring programs for citrus insects use various methods. Often two methods are combined to complement each other, such as trapping and visual checking for California red scale.

Visual Search. You can assess a pest population by counting the number of individuals present on a predetermined number of fruit or leaves (for California red scale, citrus thrips, citrus red mite) or found during a predetermined time interval (for orangeworms). Follow a general sampling scheme that you can use for several pests. For example, at five locations in the orchard or block, designate four sampling trees that you can conveniently check by walking in a circle (Figure 46). You can pick the permanent locations where you installed a trap for the California red scale (see below) or choose the locations randomly each time you visit the orchard. Inspect five fruit (for California red scale and citrus thrips) or five leaves (for citrus red mite) per tree while walking around the tree. For certain orangeworm species, monitor particular quadrants of the tree (see p. 77).

The thresholds based on these techniques, particularly on the time-search method, are conservative and can be adjusted according to your monitoring experience and the individual orchard conditions. Often a hand lens (Figure 47) and sometimes a microscope are necessary to identify a pest stage or natural enemy correctly.

Traps. Traps attract insects by scent, color, shape or ultraviolet light. They can detect an insect population, reveal its flight activity, and help you decide whether and when a pest population needs treatment. In citrus, two types of traps are available for monitoring insects: the pheromone

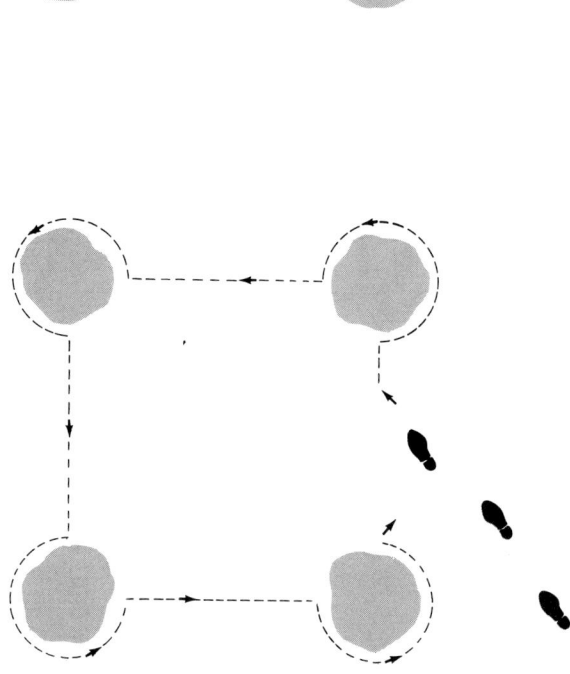

Figure 46. You can use a general monitoring scheme such as this one to inspect fruit for California red scale or citrus thrips, to sample leaves for citrus red mite, or to search for orangeworm larvae. At five locations in the orchard or block, designate four sampling trees that you can monitor conveniently by walking in a circle. See individual pest sections for further details.

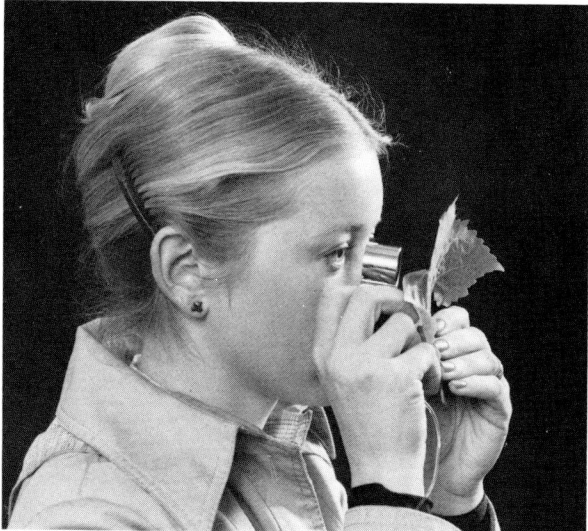

Figure 47. When using a hand lens, hold the lens close to your eye and move the object being viewed close until it is in focus.

Figure 48. You can use pheromone traps to monitor the flights of the male California red scale. A pheromone-impregnated rubber cap attracts the males, which get stuck on the sticky card. Use clear plastic to wrap the cards for counting.

trap, which uses an attractive scent, and the yellow card trap, which uses an attractive color.

Insects produce pheromones to communicate with individuals of their own species. Sex pheromones, which are released by females or males to attract the opposite sex, have found wide use in insect trapping. In citrus, pheromone traps are available to monitor the flights of the male California red scale (Figure 48). In a commonly used trap design, a rubber cap impregnated with the synthetic female pheromone slowly releases the scent, and attracts males which get stuck on a sticky card. The pheromone trap is specific for the California red scale, although other species, such as thrips and parasitic wasps, are sometimes caught. Sex pheromones for several other citrus insects, including the yellow scale, citrus and Comstock mealybug, citrus cutworm, fruittree leafroller, omnivorous leafroller, and orange tortrix, have been synthesized, but traps are commercially available only for the latter three species.

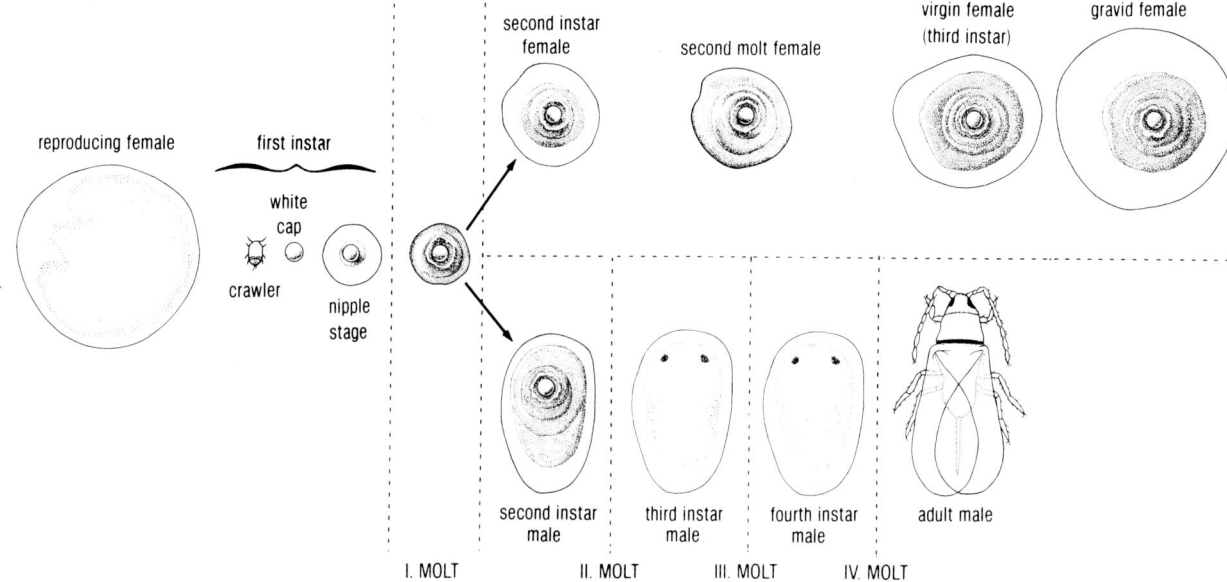

Figure 49. Development of the California red scale. Until the first molt ends, development of the female and male scales is the same. During the middle of the second instar, the female extends its round, gray margin, whereas the male develops an elongated cover. The females molt once more, the males three more times. After the second molt, the female incorporates the cast skin into the gray margin of the previous stage and begins to extend the margin further; the third instar male begins to develop eye spots and appendages (the scale cover turned over). In the fourth instar, the male appendages are completely developed. After the fourth molt, the two-winged adult male emerges. By this time, the third instar virgin female has extended its margin and is ready to be fertilized by the emerging male. The gravid female extends its margin further to reach the full size of the reproducing female. (The illustrated reproducing female is turned over to show body with posterior lobes filling the scale cover).

SCALE INSECTS

A number of species of both armored and unarmored (soft) scales occur on citrus. Because of effective biological control, soft scales are almost everywhere far less important pests than the armored scales. Several characteristics distinguish the armored scales from the soft scales. The cover of the armored scale is separated from the body of the insect, except in the adult California red scale, whereas the cover of the soft scale is the actual body wall of the insect and cannot be removed. Armored scales do not secrete honeydew, as do soft scales. Armored scales are mobile only for a day or two after hatching; when they settle to feed, they lose their legs and antennae and become sessile. The males are mobile in the adult stage. Soft scales can move around during the immature stages and retain their antennae and legs throughout their life.

The most important armored scale on citrus is the California red scale. In coastal areas and some inland districts of southern California, natural enemies can provide good control. In the San Joaquin Valley, a chemical treatment is regularly needed, but careful monitoring can limit treatments to 2-year or longer intervals. Most other scale insects are only a problem when their biological control is upset by chemical treatments, ants, or adverse environmental conditions.

California Red Scale
Aonidiella aurantii

Its wide distribution and detrimental effect on the crop and tree make the California red scale one of the most important pests in California citrus. Lemon is the most favored host, followed by grapefruit, Valencia, navel and mandarin orange. The California red scale has many other occasional host plants, such as grape, olive, rose, night-shade, eucalyptus, fruitless mulberry, and walnut. Effective parasites have greatly reduced the populations of the yellow scale, a closely related species, and the California red scale has replaced it as a major pest.

Description and Seasonal Development

The adult female has a roundish cover, which is firmly attached to the substrate when the scale is molting or reproducing. The California red scale and yellow scale can reliably be distinguished only under a microscope. In the field, however, distribution of the scales in the tree gives an indication of the species: the yellow scale is mostly found on leaves and fruit, whereas the California red scale settles on wood as well as on fruit and leaves.

The development of the California red scale is illustrated in Figure 49. The female can give birth to 100 to 150 active crawlers at a rate of two to three per day. The crawlers emerge from under the female cover in a day or two, depending on the temperature. They move about to find a suitable place to settle. During this period, they can be spread by wind, birds, or picking crews. The crawlers settle in small depressions on twigs, fruit, or leaves and start feeding by inserting their styletlike mouthparts into the plant tissue. They begin to form a circular cover by secreting cottony filaments (the white cap stage). The scale grows and forms a more solid cover, with only the center nipple of the white cap remaining (the nipple stage).

During the second instar, females and males begin to develop differently. The females molt once more and remain round. After each molt, the females incorporate the cast skin into the scale cover and extend the rim by rotating underneath the cover while excreting wax from their glands. This leaves distinct concentric rings, which reveal the stage of development. After mating, the female scale body expands and eventually fills the cover. A membrane on the underside attaches the mated female securely to the citrus leaf, twig, or fruit. At this stage, parasites can no longer attack the females.

The males molt four times before reaching maturity. The male does not rotate while extending its cover, so that an elongated cover results. Eye spots appear in the developing male about the middle of the second instar. In the third instar (the prepupal stage), appendages start to develop. During the fourth instar (the pupal stage), antennae, wings, and style are conspicuous. After the fourth molt, the small, two-winged males emerge. They do not feed and live only for about 6 hours, with the sole purpose of mating.

Virgin females attract males by releasing a pheromone. Males may crawl to nearby females or fly to other trees. They are weak flyers and cannot fly in winds over one mile per hour. Flight activity occurs at dusk when temperatures exceed 15.5°C (60°F). The males fly at a height of about 2.5 m (8 feet) and are attracted by light colored, reflecting objects, such as yellow or white trap cards.

Males of the same generation emerge and fly in search of females over a period of several weeks in what are termed *flights*. In the San Joaquin Valley, four male flights occur between March and November: the first flight occurs between mid-April and mid-May, the second flight between mid-June and late July, the third flight from early August to early September, and the fourth flight in early September, often extending into late October. The flights are synchronized with female development: at the beginning of the flight, females have completed their second molt and are in the gray adult stage, ready for mating. The flight pattern may vary; unseasonable weather during spring may advance or delay peak flight as much as two weeks. In some years, the third flight may be difficult to separate from the fourth flight because of an overlap in scale development and hot weather.

In southern California, there are generally four to seven distinct flights between April and October and sometimes November. In coastal districts, four to six male flights occur. In both areas, there are more flights than generations because each generation has two separate broods or cohorts, which sometimes overlap.

The number of generations produced by the California red scale differs with the region. In the San Joaquin Valley, about three generations appear. In coastal and interior districts of southern California, two to three generations may appear. The development of one generation requires about 1100 degree-days (°F), with a developmental threshold of 53°F (11.7°C). All stages of the scale overwinter on citrus. In the San Joaquin Valley, however, harsh winter weather often kills early developmental stages.

Damage

California red scales attack all parts of the tree, including twigs, leaves, branches and fruit, sucking plant tissue with their long, filamentous mouthparts. Fruit moderately infested with scales are downgraded at the packinghouse, and severe infestations may reduce crop production and cause fruit drop. The most serious damage the scale causes, however, is to the overall tree. Severe infestations cause a yellowing of leaves, leaf drop, and dieback of twigs and limbs. If not controlled, high populations can kill the tree, making the California red scale the most potentially injurious insect pest of California citrus. Damage to the tree is most likely to appear in late summer and early fall when scale populations are highest and moisture stress on the tree is greatest.

Management Guidelines

Biological Control. Biological control can be effective against the California red scale but depends on the climate of the growing region and good pest management. The most important natural enemies are several parasitic wasps. *Aphytis melinus* is dominant in many coastal-intermediate and interior districts of southern California. In orchards close to the coast, *Aphytis lingnanensis* is sometimes the most prevalent *Aphytis* species. *Encarsia perniciosi*, another important parasite, also occurs in coastal valleys. *Comperiella bifasciata*, an effective parasite of the yellow scale, has a host-specific race that supplements the control of the California red scale.

In the San Joaquin Valley, *Aphytis melinus* is not significant in commercial orchards, although it often provides control on backyard trees. *Comperiella* parasitizes the California red scale in commercial orchards but does not control populations; it does, however, control the yellow scale. In the Coachella Valley, the red scale occurs in a few locations where it is subject to an eradication program, and no parasites have been established. In the citrus groves of the Imperial Valley, *Aphytis* and *Comperiella* are established, but the extreme temperatures in that region do not allow it to build up to large enough numbers to provide substantial control.

Differences in the biology and appearance of the parasites help distinguish them in the field. *Aphytis melinus* lays its eggs under the scale cover *on* the scale body; the immature parasite feeds on the scale body from the outside, leaving a flat and dehydrated scale body beneath the scale cover. The yellow adult may emerge through a small, round exit hole in the scale cover or push out between the scale cover and plant substrate, so that parasitized scales often slough off. *Comperiella* lays its eggs *inside* the scale body, and the developing parasite feeds inside the scale body, leaving an empty, bloated shell of scale skin. The black adult emerges through a large, irregular exit hole. *Encarsia perniciosi* is an internal parasite that develops similarly to *Comperiella*. Like *Comperiella*, it is dark colored, but it is smaller and does not have white stripes on its head.

Aphytis controls California red scale both by laying its eggs onto the scale body and by feeding on it (host feeding). Although *Aphytis* parasitizes mainly the third instar virgin females, the limitation is largely offset by its short life cycle (about two generations per each scale generation) and by the host feeding. To host feed, the *Aphytis* adult punctures the scale body with its ovipositor, then ingests the exuding fluid. *Comperiella* adults do not feed on the scale host and have only one generation for each scale generation.

Several predators feed on the California red scale, although they are not nearly as important as the parasites. *Lindorus lophanthae* is a small, metallic black ladybeetle, which at times becomes numerous in some coastal and interior groves of southern California with high scale infestations. *Chilocorus stigma*, another ladybeetle, and many other general predators will feed on the red scale but usually do not have an impact on its control.

Careful pest management can aid the natural control of the California red scale in the orchard. Most chemical treatments for California red scale or other pests are destructive to natural enemies. When possible, choose those chemicals that are least disruptive to biological control and use pesticides only when needed; follow a monitoring program to indicate when and if scale treatment is necessary.

Ant colonies disrupt biological control because ants protect scales against attack by parasites (see *Ants*, p. 99). Excessive dust on leaves and fruit also interferes with parasitism. Minimize dust by reducing traffic in the orchard and by oiling orchard roads.

Monitoring and Control. You can monitor the California red scale by visually checking fruit, twigs, and leaves for scales and by using a pheromone trap to monitor the male flights. The pheromone trap reduces the amount of visual inspection necessary, although it does not replace the checking completely.

Visual Inspection. Inspect fruit at least once before harvest and preferably three to four times a year. When using traps, select fruit sampling trees in the vicinity of each trapping site, not more than two rows away in each direction. Follow a sampling scheme, such as the one outlined under *Monitoring Methods* (page 55) and check at least 20 fruit at one site. Do not injure or remove the fruit. Count all fruit that have *10 or more* live scales as infested and record your observations. The level of tolerable infestation depends somewhat on market conditions, crop yield, presence of natural enemies, and other pest problems. See regional programs for details.

Pheromone Traps. Maintaining pheromone traps can serve two purposes: A trapping program can help you assess the damage potential of a population; there is a good correlation between the number of males caught and the proportion of infested fruit. Thus the program is valuable in deciding whether a chemical treatment is warranted. Yellow card traps also provide an indication of *Aphytis* activity and thus help in deciding whether a release of additional parasites is beneficial. Trapping can locate a scale population, particularly in a new or young orchard, long before a visual check would detect an insipient infestation; thus action can be taken before damage occurs. Over time, trap records provide valuable information about relative density and distribution of the scale population and about the effectiveness of control measures.

Certain problems have to be considered when using male flights to assess potential scale problems. Because flights are not necessary for mating, trap counts are lower at times than the actual numbers of males present. In addition, cool or windy weather, as often occurs on the coast, may reduce the effectiveness of the pheromone traps.

The several types of pheromone traps available are equally effective in catching California red scale males. The most commonly used design is the paper clip card trap, consisting of a yellow or white card, to which a rubber cap impregnated with the sex pheromone is attached with a paper clip (Figure 50). Where *Aphytis* occurs, the yellow trap card also attracts the male parasite, revealing the parasite's presence and relative density. In the San Joaquin Valley, where parasites are not effective, a white card is generally used. You can obtain plastic cards (7.6 x 12.7 cm or 3 x 5 inches) from plastic suppliers and the pheromone caps from the Zoecon Corporation or chemical supply houses. Punch a hole at one end of each card and coat it on both sides with Tangletrap® or Stikem Special®. To assemble the trap, insert an unfolded paper clip through the hole and attach the pheromone cap with the large end pointing downward. When removing the cards from the traps, envelop them in clear plastic sheets or sandwich bags for easy transport and reading. You can also use a carton trap for monitoring the flights. Place the pheromone cap inside a half-pint ice cream carton, replace the lid with a fine mesh, and attach a sticky, yellow or

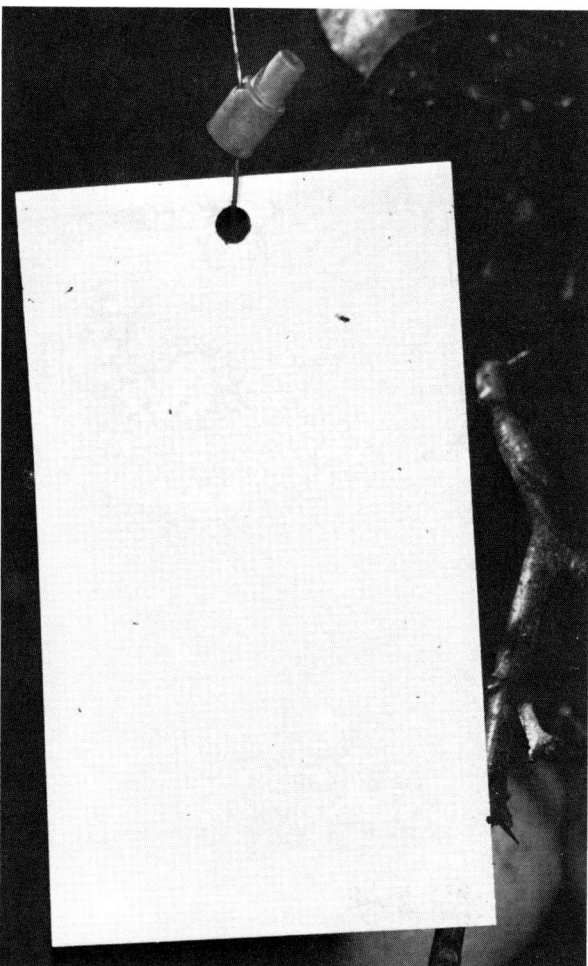

Figure 50. The most commonly used trap design for monitoring the flights of the male California red scale is the paper clip card trap. Coat a white or yellow 3- x 5-inch card with a sticky substance and attach the rubber cap impregnated with sex pheromone with a paper clip.

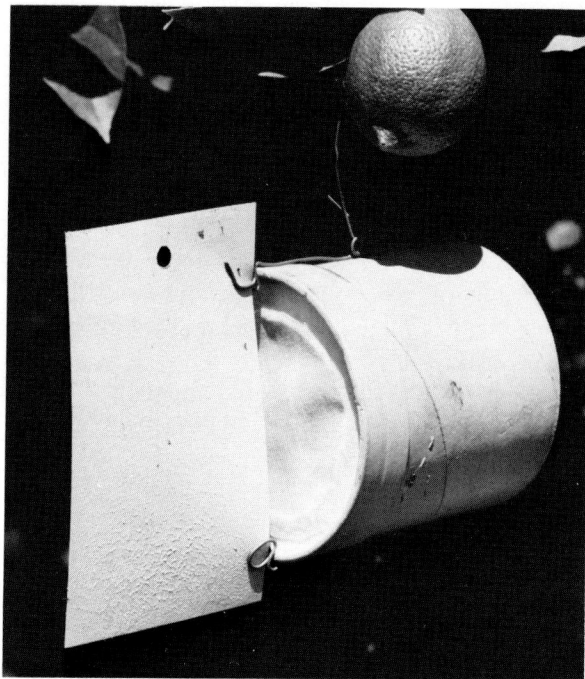

Figure 51. You can also monitor the California red scale with a carton trap. Place the pheromone cap inside a half-pint ice cream carton, replace the lid with a fine mesh, and attach a sticky, 3- x 5-inch card across the opening.

Figure 52. A tent trap for monitoring the California red scale is commercially available.

white 3- x 5-inch card across the opening (Figure 51). A tent trap consisting of a folded cardboard card and the pheromone cap is commercially available (Figure 52). For all trap types, change the pheromone caps every 4 weeks (every 6 weeks in cool regions).

Hang the traps 6 to 8 feet high in the northeast quadrant of trees slightly inside the canopy. Install a minimum of one trap per ha. (2.5 acres), space traps at the same distance from each other if possible, and choose trees at least four rows into the orchard. Check traps at regular intervals; the frequency depends on your needs. In the San Joaquin Valley, where the trapping program is used for assessing the need for treatment, a trap check once a week during the flight period is recommended. Where traps are used mainly as a survey and detection tool, as in southern California, less frequent checking is sufficient.

You can count the males under a dissecting microscope at about 7x, although with experience you can recognize the males with the naked eye. Identification of the males is important because other small insects, such as thrips, parasitic wasps, whiteflies, and midges, may also be caught by these traps. Adult male scales are light reddish brown, have a dark, narrow band across the back, close-set eye spots, and threadlike antennae, which are longer than the short, broad body; the grayish wings may or may not be visible on the sticky card. *Aphytis*, the species most closely resembling the male scale, is yellow, has short, knobby antennae, and has no band on its back.

The trap counts can be used to make treatment decisions where flights are fairly distinct, as in the San Joaquin Valley. In southern California, where flights are more erratic, treatment decisions are based mainly on fruit counts. The choice of materials depends on the severity of the infestation, the natural enemy spectrum fauna present, and climatic conditions. Synthetic organic materials, such as organophosphates and carbamates, kill a higher percentage of California red scale but are also more harmful to beneficial species than spray oils. Oils, either as dilute spray or concentrate spray, provide adequate control in many situations, but certain precautions have to be taken. For all treatments, thorough coverage, especially of the center tree tops, is important for effective control. For details, see *Pesticide Application*, p. 30. The timing of a treatment is critical for good kill; oils and synthetic materials are most effective on crawlers and early instars on twigs and leaves.

You can accumulate degree-days (see p. 14) to predict the appearance of flights and time the pheromone trapping program most efficiently. For example, about 2200°D (two generation times) pass between the second and fourth flight. You can keep records of accumulated degree-days based on daily temperatures in your own orchard in tabular form or on your home computer. Degree-day information is also available from public sources. The University of California IPM computer processes temperature data from various locations and develops degree-day data for the California red scale. A number of growers are currently using these data on an experimental basis. Expanding the degree-day program to all citrus-growing areas is planned for the near future; check with your farm advisor to see if the program is available in your district.

Management Program for Navel Oranges in the San Joaquin Valley. In the San Joaquin Valley, biological control for the California red scale is not effective at present; thus control actions, when needed, consist of pesticide applications. Most growers use organic synthetic chemicals; however, oil sprays can be used effectively without harm to the trees when proper precautions are taken.

To assess the need for treatment, monitor the second (June/July) and fourth (September/October) flights with pheromone traps. Male catches usually increase markedly between the second and fourth peak flights (10 to 20 times or more) except where the second flight is very large.

To start your trapping program, follow the schedule provided by your farm advisor's office, or install a few survey traps in late May and check them twice a week; when the first males are caught, install the remaining traps. To determine the length of the flight, continue checking one trap card twice weekly and plot the number of males/trap/day on a graph. You can easily see the beginning and the end of a flight (Figure 53). Count the days between a definite increase and a consistently low level as the flight period. Figure 53 represents a typical second flight; the pattern of the fourth flight is more difficult to interpret. If you are in doubt about the length of a flight, check with your farm advisor's office.

Change the rest of the cards weekly and count the trapped male scales. If less than 200 males are caught on both sides of a card, count each individual male with a tally counter and record the total on a form (Figure 54). If you estimate that more than 200 males are present, use one of several templates, depending on the type of trap. You can copy the templates printed on the inside book covers onto a clear plastic sheet, or you can make your own. Divide a clear plastic sheet the size of the trap card into 60 squares and designate 20% (12 squares) for counting; use the distribution given inside the front cover for the paper clip trap card and the square arrangement inside the back cover for the carton trap card. No template design is available for the tent trap card. Count the number of males within the clear squares and multiply by 5 to obtain the approximate total number of males on the card. Carry out the procedure for the front and back sides of the cards, and record your trap counts. Note the total number of males from all trap cards in the block each time you change the cards. At the end of the flight, add up the figures and divide the total by the number of traps in the block to get the number of males/trap.

The number of males/trap for a flight has been correlated with the percentage of infested navel orange fruit

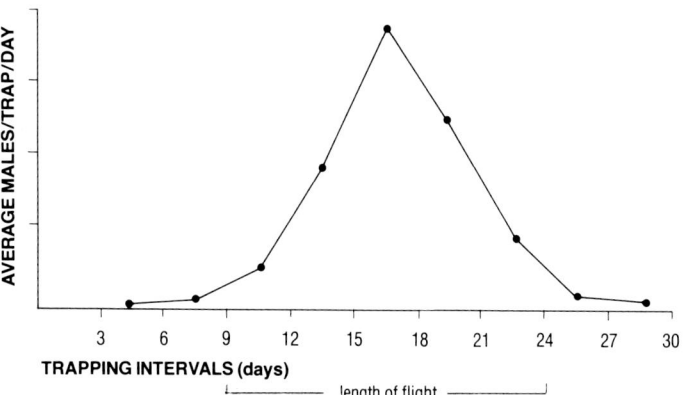

Figure 53. To determine the length of a flight of the California red scale, plot the average number of males/trap/day over the trapping period. The days between a definite increase and a period of consistently low catches give the length of the flight.

at maturity in the San Joaquin Valley. Based on your trap counts, you can estimate from Table 9 what level of fruit infestation to expect, and you can choose a threshold percentage acceptable to you. The fourth flight is used to make treatment decisions; experience has shown that based on the trap catches of the fourth flight, the infestation level at harvest one year later will be five to ten times the level of the current year's crop if no treatment is made. The following guidelines have been established for navel oranges in the San Joaquin Valley:

- If counts of the fourth flight indicate the need for treatment, the best time to spray is early the following June when a high percentage of the scale population is in a susceptible stage and crawlers have not yet moved onto the new fruit.
- If the counts of the fourth flight indicate no treatment is needed, recheck the second flight of the following year. If the infestation exceeds the tolerable projected level, treat immediately.

Supplement the trapping program with a visual check of fruit (page 59) once during August, October, and December and record your counts on your summary sheet (Figure 45).

Table 9. Predicted Fruit Infestation Levels Based on California Red Scale Male Trap Catches in the San Joaquin Valley.[1]

MALES PER TRAP IN THE FLIGHT PERIOD			PERCENTAGE OF FRUIT INFESTED	
First Flight (Apr/May)	Second Flight (Jun/Jul)	Fourth Flight (Sep/Oct)	one or more scale per fruit	11 or more scales per fruit
	557	3,621	2	0.7
	2,135	7,405	4	1.4
	3,646	11,275	6	2.1
	5,501	15,287	8	2.7
	7,333	19,299	10	3.4
39	9,185	23,462	12	4.0
95	11,070	27,731	14	4.7
136	13,000	32,217	16	5.3
172	15,000	36,727	18	6.0
193	17,000	41,080	20	6.6
269	19,000	46,428	22	7.3
327	21,210	51,530	24	8.0
383	23,406	57,000	26	8.6
431	25,663	62,504	28	9.3
489	28,233	68,449	30	9.9
546	30,644	75,000	32	10.6
605	33,417	81,472	34	11.2
663	36,316	89,000	36	11.9
723	39,350	97,000	38	12.6
784	42,533	106,083	40	13.2

1. Fruit have not generally been scored as infested by packinghouses unless they have patches of more than 10 scales per fruit.

This management program has been used for several years in commercial navel orange groves; in most cases, treatment was necessary only every two years. In a few cases, treatment was not needed for three years. For insecticide recommendations, see the latest University of California *Treatment Guide for California Citrus Crops*, Leaflet 2903, Agricultural Sciences Publications.

Management Program for Other Cultivars in the San Joaquin Valley. The same monitoring program used for navels can be used for Valencias. Table 9, however, is based on infestation levels of navels, and threshold levels may be lower for Valencias if they are harvested after a new generation of California red scale has emerged in late spring or summer.

No management guidelines are available for lemon. Most lemons require chemical treatment every year, and one correctly applied application in June provides good control. As in other programs, trapping can be used to check the effectiveness of a treatment.

Management Program for Lemon in Coastal-Intermediate Districts. Because of their repeated growth flushes, coastal lemon trees favor the buildup of of California red scale. Biological control, however, can be significant, and treatment decisions and choice of materials should take the level of parasitism into account. The correlation between trap counts and fruit infestation is not as good as on navel oranges in the San Joaquin Valley, so visual inspection of the fruit is currently the most precise way to determine need for treatment.

Check the fruit at harvest and preferably once a month on the tree during the growing season (see p. 62). The economic threshold for fruit infestation of lemons varies substantially with the market and crop yield; it has been found to average 3 to 5% fruit infested with more than 10 scales in late fall.

Experimental programs using pheromone trap counts are being tested in some areas. Treatment decisions in such programs are based on the July flight. If you use this program, maintain the pheromone traps from about late June through late July; farm advisor or PCA newletters often report the precise trapping time. Check one trap twice a week to determine the peak flight and the remaining traps once a week. Count the trapped males and calculate the number of males/trap/day for the 2-week peak activity (see program for navel oranges). Use only the middle 14 days of the flight because at present it is difficult to determine the beginning and end of the flight. Thresholds for different areas and orchard conditions are being developed; check with your local farm advisor for the latest information.

Maintain a few traps in key areas to monitor the population level for the rest of the season (through October). If a treatment or parasite release was successful, trap counts should be greatly reduced for the remainder of the year.

When control actions are needed, it is often sufficient to release additional parasites, which are available from certain insectaries. A parasite release is most effective at or before the beginning of a flight when most scales are in a susceptible stage. If chemical treatment is necessary to bring down a high population quickly, oil is the preferred material because it is less hazardous to beneficial insects, particularly when applied as a low-volume spray. Parathion is more selective than other synthetic materials used for scale control and provides good control. For insecticide recommendations see the latest University of California *Treatment Guide for California Citrus Crops*.

Management Program for Valencia Oranges in Interior and Coastal-Intermediate Districts of Southern California. The California red scale on Valencia oranges in interior and coastal-intermediate areas is generally under good biological control unless upset by dust, ants, or chemical treatments, mainly those for citrus thrips. Valencia oranges can be monitored in the same way lemons are monitored. Check fruit periodically for the level of scale infestation, following a sampling scheme such as the one outlined under *Visual Inspection* (page 59). Use pheromone traps to survey the relative population level of the California red scale and its parasites in different orchards, to detect an insipient scale population in a young or new orchard, or to time a parasite release. The July flight is usually most reliable for monitoring. A strong correlation between trap counts and fruit infestation is difficult to establish because each crop of Valencias is usually exposed to scale infestation for more than one year. The acceptable level of fruit infestation, resulting in no culling at the packinghouse, is between 5% and 10% fruit infested (10 or more scales) in the field.

When control is needed, a release of parasites or an oil application are the preferred options. A release of parasites is most effective before or at the beginning of a flight. Farm advisor newsletters often announce the beginning of flights. Several insectaries in southern California raise the major parasite, *Aphytis melinus*. Check with your farm advisor for information on these sources. Oil sprays have traditionally been an important part of the integrated control program in southern California. They provide adequate control and are least disruptive to natural enemies of the California red scale and other pests, thus avoiding the pest resurgence and secondary pest outbreaks often associated with the use of synthetic organic materials.

Management program for Citrus in Desert Districts of Southern California. In the Coachella Valley, the California red scale is under an eradication program. In the Imperial Valley, where the scale is more widespread, parasites are established but are largely ineffective. When control actions for the California red scale are needed, synthetic organic materials are used, since oil sprays are likely to damage the trees in the hot, dry climate of the desert.

ORCHARD _____ BLOCK _____

TOTAL NUMBER OF TRAPS _____

COMMENTS _____

Date	CRS males caught					
	Trap 1	Trap 2	Trap 3	Trap 4	Trap 5	TOTAL

Length of flight (Total Days) ☐   Total males for flight ☐

Total males/trap ☐

Figure 54. Record sheet for trap catches of the California red scale (CRS) in the San Joaquin Valley. Record the total number of males from all traps for the block each time you change the trap cards. At the end of the flight, add the figures for the total number of males for the flight. Divide the total by the number of traps per block to get the total number of males/trap. Use this figure to project fruit infestation at the end of the year (Table 9).

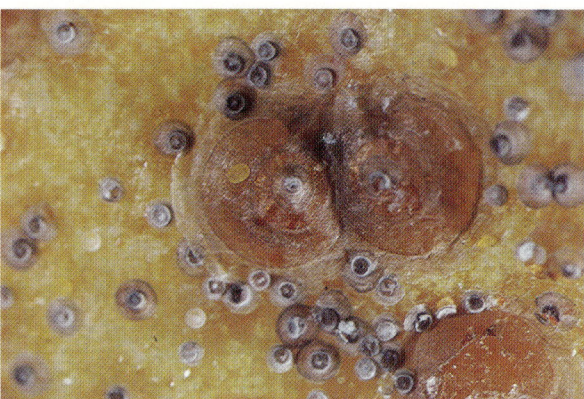
Various developmental stages are usually present in a colony of California red scale. Shown here are large, round adult females, a small yellow crawler on a female, and young scales in the nipple stage.

Aphytis melinus (left), an important parasite of the California red scale, has a yellowish body and short, knobby antennae. *Comperiella bifasciata* (right), another parasite of the California red scale, is distinct because of its black color, white head stripes, and fork-tailed wings.

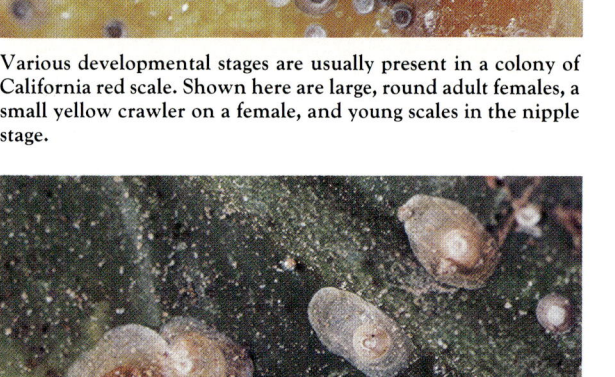
During the mid-second instar, the males of the California red scale (top right) start developing their elongated scale covers. A parasite has emerged from the exit hole in the round female at left.

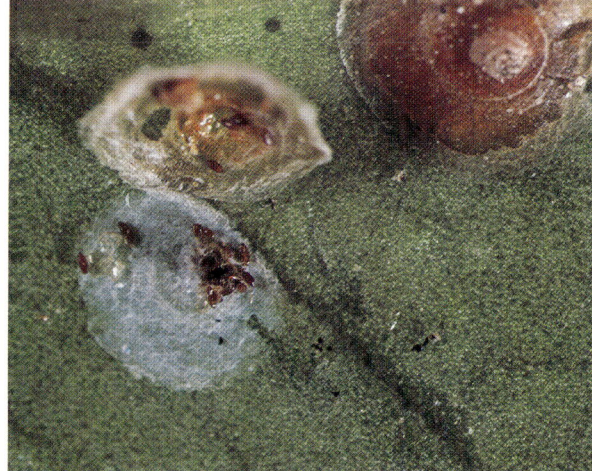
California red scales that have been parasitized by *Aphytis* have a small, round exit hole. When you lift the cover, you can usually see meconium pellets and the dried, flat skin of the scale body.

The adult male of the California red scale is yellow to light brown, has a dark band across its back and long antennae.

California red scales parasitized by *Comperiella* have a large, irregular exit hole and meconia inside an empty, bloated scale skin.

The California red scale may infest fruits, leaves, or twigs.

Severe infestation with California red scale—here of a grapefruit tree—results in leaf drop and dieback of twigs and branches, usually starting in the northeast part of the tree.

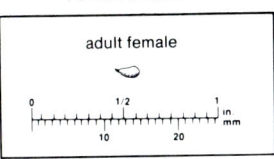

The purple scale cover resembles a mussel shell and has a purplish tinge. The scale is generally under good biological control; look for the parasite exit holes in scale covers.

Purple Scale
Lepidosaphes beckii

In California, the purple scale is an occasional pest only in certain coastal areas. Mild climate and humid conditions favor its buildup, but parasites usually provide good control. Unlike many other scale insects infesting citrus, the purple scale is mainly a pest of citrus and has few other host plants.

Description and Seasonal Development

The adult female cover resembles a mussel shell. The male cover is shorter and much narrower than the female cover. When mature, the winged male emerges from under the cover. After mating, the female lays 40 to 80 eggs under her cover.

The major parasite of the purple scale is *Aphytis lepidosaphes*.

The crawlers settle on branches, twigs, leaves or fruit and begin to form their covers. Until about half grown, they are covered with a mass of waxy threads. The purple scale prefers the cooler, shady part of the trees; temperatures above 29°C (80°F) and a humidity below 70% greatly reduce a population. Two generations are produced between May and October; a third generation may be partially completed before the cold weather starts.

Damage

The purple scale attacks all parts of the tree. Yellowish haloes develop on leaves around sites where the scales have been feeding; on young fruit, the feeding sites remain green. When populations are high, purple scales can cause defoliation and kill twigs. Damage is usually most severe in limited patches on the lower north side of trees.

Management Guidelines

Biological Control. Several natural enemies attack the purple scale and are generally able to keep it under control. By far the most effective is a parasitic wasp, *Aphytis lepidosaphes*, which is now generally distributed in the citrus-growing areas where the purple scale occurs. This parasite develops externally on the body of immature scales. Up to eight individuals may reach maturity on a single scale, although most scales yield only two parasites. The female wasps also kill many scales by host-feeding on the body fluids of the scales.

Several predators, among them the twice-stabbed lady beetle, *Chilocorus* spp., and the Australian ladybeetle, *Lindorus lophanthae*, have been important in reducing peak populations, but apparently are not able to maintain purple scale below economic levels without assistance from the parasite.

Monitoring and Control. The natural control of the purple scale may require supplementary treatment at times. Purple scales are more likely to build up on dusty trees, such as those next to dirt roads. If a treatment is needed, it may be sufficient to spot-treat the dusty trees with an oil spray or even wash them with water. If the infestation is heavier and more generally distributed in the orchard, you may want to treat the entire orchard and depend on the survival of *Aphytis* or its migration back from other orchards. Oil sprays for the California red scale also control the purple scale.

Citricola Scale
Coccus pseudomagnoliarum

The citricola scale is a pest of citrus only in the San Joaquin Valley, where chemical treatments for the California red scale and other pests currently control this scale. In southern California the citricola scale is now rare, apparently because of effective parasitism. Besides citrus, it may also infest elm, walnut, night-shade, pomegranate, and hackberry.

Description and Seasonal Development

The citricola scale may at times be confused with the brown soft scale, but color pigmentation and number of generations distinguish the two species. The citricola scale is a mottled dark brown when young and gray at maturity; the brown soft scale is a mottled yellow brown when young and dark brown at maturity. The citricola scale has one generation a year, so all individuals are approximately in the same developmental stage; the brown soft scale has three to five overlapping generations a year, so different developmental stages are present at one time. The citricola scale has a high reproductive potential: the females may lay 1000 to 1500 eggs over a period of 1 to 2 months, and the eggs hatch after 2 to 3 days. The first crawlers may appear in late April.

The crawlers settle mostly on the underside of leaves; in severe infestations, they also settle on the upper leaf surface and on twigs, but rarely on fruit. While on leaves, the young scales are flat and almost translucent. The scales grow slowly during the summer and molt twice. By November, they become much darker and start migrating to twigs; this migration peaks in February and March. On the twigs, the scales develop rapidly, become mottled gray, and mature by late April or early May. Males are usually not present.

Damage

Citricola scales reduce tree vigor by sucking plant juices from leaves and twigs. A severe infestation may kill twigs and reduce flowering and fruit set. Sooty mold grows on the honeydew excreted by the insects. The mold does not damage the fruit, but it is often so heavy that it does not wash off and results in downgrading of the fruit. A heavy coating of sooty mold on leaves interferes with photosynthesis. Although the most active feeding occurs during the spring, the resulting damage appears mainly in late summer.

Management Guidelines

The natural enemies of the brown soft scale (*Metaphycus luteolus*) and the black scale (*Metaphycus helvolus*) apparently also control the citricola scale in southern California. Although these parasites are established in the San Joaquin Valley, biological control of the citricola scale is not effective.

You can check for citricola scale when monitoring other scales, but look especially closely during the spring and late summer. In the spring and early summer, you can find the adults on twigs. During the summer and fall, immature scales feed on the underside of leaves. Honeydew and sooty mold on leaves or fruit may indicate an infestation by the citricola scale, although heavy populations of aphids or other soft scales also produce these symptoms in mid-spring. Be sure to distinguish the citricola scale from the brown soft scale. Check about 50 leaves per tree in the northeast quadrant, where citricola scale populations tend to be heaviest. If you find an average of more than one-half to one scale per leaf, a treatment is warranted.

In orchards in the San Joaquin Valley not regularly treated for the California red scale, the citricola scale may

The citricola scale has dark brown pigmentation when young but becomes grayish at maturity.

A heavy infestation of citricola scale produces copious amounts of honeydew, which covers leaves and fruit and favors the growth of sooty mold.

Psocids are common in citrus orchards in the spring, when they feed on honeydew and sooty mold resulting from citricola and other scale infestations. Adults are about 4 mm (¼ inch) long. An adult and a small wingless nymph are shown here.

Some psocid species lay their eggs in clusters and cover them with silken threads.

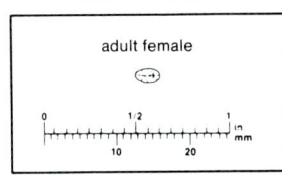
CITRICOLA SCALE
adult female

Young brown soft scales are yellowish, mottled, and rounded.

Mature brown soft scales are dark brown and somewhat flattened. A mature parasite has emerged from the scale at the bottom.

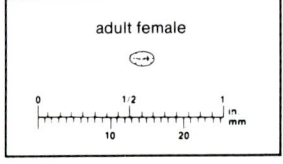

Brown Soft Scale
Coccus hesperidum

In California, the brown soft scale is commonly found on citrus in all growing areas and also on avocado and other subtropical fruit trees. Effective parasites usually prevent it from becoming an economic pest. A moderate to heavy infestation may occur when biological control is disrupted.

Description and Seasonal Development

The brown soft scale may be confused with the citricola scale but several characteristics distinguish the two species (see *Citricola Scale*). Female brown soft scales lay a few eggs at a time during the summer. The eggs hatch almost immediately, and the crawlers start to feed. The young scales move around until they are about half grown. The young molt twice and reach maturity on leaves or twigs; they rarely move onto the fruit. No males have been observed in California.

The brown soft scale produces three to five overlapping generations a year. Populations are usually highest from midsummer to early fall.

Damage

The brown soft scale causes damage similar to that of other soft scales. Its feeding reduces tree vigor, kills twigs, and reduces yields. Sooty mold grows on the excreted honeydew and may affect the fruit grade. The honeydew attracts ants, which interfere with the biological control of the scales.

Management Guidelines

Several natural enemies can provide control of the brown soft scale. The most effective enemy in all citrus districts is a parasitic wasp, *Metaphycus luteolus*. It destroys the scale in the early instars before the host can produce offspring or can cause substantial injury. A ladybeetle, *Chilocorus cacti*, preys on brown soft scales.

Monitor the brown soft scale during the warm part of the year, especially from June through October, when disruption of biological control may be a problem. Check the level of parasitism by looking for parasite exit holes or for developing parasites within the scale body.

Management of the brown soft scale focuses on preserving the effective natural enemies. When controlling thrips or orangeworms in the spring, avoid the repeated use of organophosphates. A buildup of scales and honeydew in early summer is the first sign of a disruption of biological control. The natural enemies usually recover and reduce the scale population within several months.

Ants can also disrupt biological control. Ants feed on the excreted honeydew and protect the scales from attack

reach damaging levels. A high population may require treatment before bloom. If the population is low, you can postpone treatment until late summer or fall when the scales are small and easier to control. A well-applied treatment often gives 3 to 5 years of control. For insecticide recommendations, see the latest University of California *Treatment Guide for California Citrus Crops*.

by the parasite. For the control of ants, see the Ants (page 99).

Individual treatment for brown soft scale is usually not necessary. Where natural enemies do not control the scales, an oil spray may be applied; a spot treatment of the infested trees is usually sufficient. Many materials used for the control of the California red scale also control the brown soft scale; parathion, however, does not control this scale.

Black Scale
Saissetia oleae

The black scale attacks citrus as well as other crops and ornamentals including pepper and olive trees, oleander and nightshade. The black scale was once a major pest, but today introduced parasites provide substantial control. This scale does not occur on citrus in the San Joaquin Valley except for occasional infestations, mostly on grapefruit.

Description and Seasonal Development

In the interior areas of southern California, the black scale produces one generation; in coastal districts, it usually has two overlapping generations. The females, which can reproduce without mating, lay an average of 2000 eggs over a period of 2 to 3 months, mainly during May and June. The crawlers move about for some time before settling mostly on both sides of leaves. In the late second instar a ridge develops on the scale's back and later expands into an H shape. The scale becomes dark mottled gray and leathery, a phase referred to as the "rubber stage." Sometime after the second molt, the young scales migrate to twigs, where they grow rapidly and become nearly circular. Once egg laying starts, the scales become darker and harder, and the H-shaped ridge often disappears. Male black scales are rarely seen.

Damage

As with other scales, the feeding of the black scale reduces tree vigor, eventually leading to leaf or fruit drop and twig dieback. Fruit may be covered with the excreted honeydew, which supports the growth of sooty mold.

Management Guidelines

Several predators and parasites have been introduced against the black scale, and today a parasitic wasp, *Metaphycus helvolus*, provides substantial control in southern California. This tiny, amber-colored wasp lays its eggs within the scale body. Adult female parasites also

The black scale develops an H-shaped ridge on its cover in the late second instar.

A parasitic wasp, *Metaphycus helvolus*, provides substantial control of the black scale in southern California. Look for the round exit holes in the scale covers.

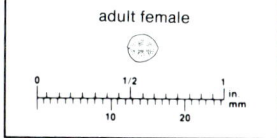

feed on the body fluids of the scale, often causing higher mortality through host feeding than through parasitism. Usually one parasite develops in second or third instar scales, although occasionally two or three wasps emerge from one scale. The effectiveness of the parasite depends on the region, climate, and conditions in the orchard. *Metaphycus* survives better in the coastal areas, where overlapping scale generations provide susceptible stages for a longer time than in the interior regions. Warm winter weather also helps the parasite because the scale development is less synchronized. On the other hand, cold weather and other adverse conditions often drastically reduce *Metaphycus* populations. A few insectaries raise *Metaphycus* to supplement existing populations in the

The cottonycushion scale is easily recognized by the white, fluted egg sac of the adult female. The vedalia beetle is an effective predator of the cottonycushion scale (adult vedalia is in foreground, its eggs and larva are on the scale).

This pupal case of vedalia shows the red coloration of the developing adult inside; often only the empty pupal case can be seen attached to leaves on the outside canopy.

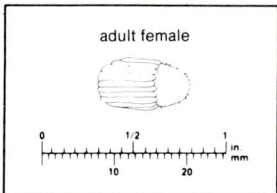

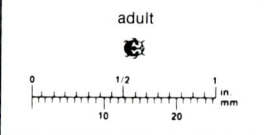

field. Releases are best made in late summer or early fall. Check with your farm advisor's office for names of insectaries and their services.

If ants are present, biological control may be severely hampered. Ants feed on honeydew and protect the scales from attack by natural enemies (for details see Ants). *Metaphycus* is particularly sensitive to disruption because it is slow in depositing its eggs and spends some time feeding on the body fluid leaking from the oviposition puncture.

The black scale usually appears in damaging numbers only when parasite activity has been upset. The scale is more often a problem in intermediate and interior districts than in coastal areas. Watch for the newly settled scales in late June or early July. An oil treatment is generally sufficient for control and is most effective when applied before the H formation becomes apparent on the young scale covers in September. For recommendations and precautions when using oil sprays see the latest University of California *Treatment Guide for California Citrus Crops*.

Cottonycushion Scale
Icerya purchasi

The cottonycushion scale was a major pest in the 1880s, threatening the citrus industry of southern California. Efforts in controlling this pest resulted in one of the earliest and most impressive examples of biological control. Today, light infestations occur every summer in the San Joaquin Valley due to the temporary destruction of the natural enemies by insecticide treatments.

The cottonycushion scale also occurs on a wide variety of fruit and nut trees and ornamentals.

Description and Seasonal Development

The most distinguishing feature of the adult female is the fluted, cottony egg sac. The female secretes this sac and lays from 600 to 800 eggs into it. The eggs hatch in a few days in the summer, but take up to 2 months in the winter.

The newly hatched nymphs are red with dark legs and antennae. First and second instar nymphs settle on twigs and leaves, usually along the veins. Third instars and adults are mainly found on branches and the trunk; they rarely move onto fruit. The third instar is covered with a thick, cottony secretion, which disappears after the third molt. The adult females settle and begin to form the white, elongated egg sac.

Males and females develop similarly up to the third instar, when the male nymph seeks a secluded place on the tree or ground to change into a winged adult inside a loose cocoon. Males are rare, and females can reproduce without mating. Cottonycushion scales produce three generations a year.

Damage

The cottonycushion scales extract plant sap from leaves, twigs, and branches, thus reducing tree vigor. When an infestation is extremely heavy, it may result in leaf and fruit drop and dieback of twigs and small branches. The secreted honeydew and subsequent growth of sooty mold covers the fruit but washes off during the processing at the packinghouse.

Management Guidelines

Biological Control. In most groves, two natural enemies effectively control the cottonycushion scale. The most famous is the vedalia beetle, *Rodolia cardinalis*, which was introduced from Australia in the early 1890s. Both adult and larva feed on all stages of the scale. The female beetle lays its eggs underneath the scale or attaches them to the egg sac, sometimes singly but usually in groups of three or more. The young larvae move into the egg mass from below and feed on the eggs. Later, the larvae feed on all scale stages.

A parasitic fly, *Cryptochaetum iceryae*, was the first natural enemy introduced from Australia. Because of the spectacular success of vedalia, it was largely ignored, but at times it is more effective than the beetle, especially in the coastal areas. The fly deposits its eggs inside the scale body. The larvae feed on the scale body and pupate within the remains of the scale. The fly produces up to eight generations per year.

Monitoring and Control. If you encounter cottonycushion scales in your orchard, look for the vedalia beetle. You may see its red eggs on the white egg sacs of the scales or find its empty pupal cases attached to outside leaves. If a few vedalia are present, they usually build up to sufficient numbers to control an infestation. Even if you do not find any vedalia, they may move in later from other orchards in search of scales. You can also recolonize vedalia by collecting them from other orchards with a cottonycushion scale infestation.

In the San Joaquin Valley, cottonycushion scales may flare up after treatment for thrips and orangeworms in the spring, but the vedalia usually recovers and controls any scale infestations by late summer. An insecticide treatment is rarely warranted.

In coastal areas, you can often observe *Cryptochaetum* parasitizing the cottonycushion scales. The emerging parasite leaves an exit hole in the mummified scale. Ants are attracted to the honeydew excreted by the cottonycushion scale but do not interfere with its biological control.

The parasitic fly, *Cryptochaetum iceryae*, is effective in coastal areas. Here its pupal case is attached to the remains of a cottonycushion scale.

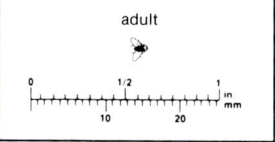

CITRUS THRIPS LARVA

- about 99% on new leaves and small fruit

- maturing larvae broadly oval shaped, light amber in color, very active; very small spines or hairs, hardly visible with hand lens

- adults extremely active; abdomen rounded, light orange yellow

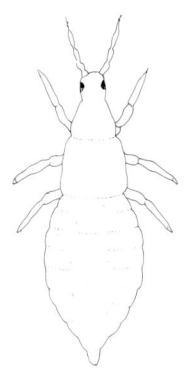

FLOWER THRIPS LARVA

- about 99% in blossoms, disperse after petal fall

- larvae slender, cigar shaped; pale yellowish to white, slow moving; spines or hairs visible with hand lens

- adults relatively sluggish; abdomen straight, cigar-shaped; straw colored or dark brown

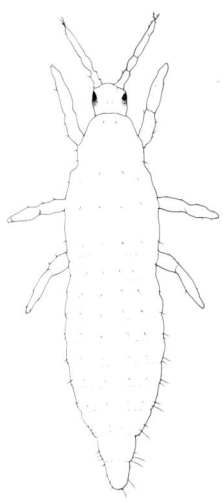

Figure 55. Distinguishing characteristics of the citrus thrips and the flower thrips (second instar larvae).

The citrus thrips (left) is a tiny, yellowish insect with fringed wings. The female (shown) has a much broader abdomen than the male. The flower thrips (right), which often feeds on citrus flowers, should not be confused with the citrus thrips; it is larger, darker, and more slender than the citrus thrips.

Citrus Thrips
Scirtothrips citri

The citrus thrips is most often a problem in the San Joaquin Valley and desert, although in recent years it has also become a pest in coastal and interior districts. It can attack all citrus cultivars but is of greatest economic importance on navel oranges in the San Joaquin Valley and on lemons in coastal areas.

Description and Seasonal Development

The adults of the citrus thrips are small, orange-yellow insects with fringed wings. It is important to distinguish citrus thrips from flower thrips, which feed on flower parts but do not damage citrus (Figure 55). Shortly after petal fall, you may see immature flower thrips moving about young fruit, but they soon pupate and the adults disperse to other plants. The predaceous sixspotted thrips may also occur on citrus, although normally not around petal fall, when most citrus thrips monitoring is done. Three dark spots on each forewing distinguish the adults of the sixspotted thrips and the citrus thrips; immatures are more difficult to identify.

Citrus thrips seek warmth and light, although they avoid extreme heat. The adults are capable fliers and leave a site when disturbed or when searching for better food. Citrus thrips prefer to feed on young plant tissue such as new leaves, green twigs, and small fruit; they do not feed on the flowers. In the San Joaquin Valley and the desert districts, citrus thrips are most abundant in the spring and fall on the new growth flushes. On lemon in coastal-intermediate or interior districts, the thrips are a problem from June through October when most new growth occurs.

During the spring and summer, the citrus thrips female lays 25 to 75 eggs in new leaf tissue, young fruit, or green twigs; in the fall, the overwintering eggs are laid mostly in the last growth flush of the season. The overwintering eggs hatch about the time of the new spring growth in March. During hatching, many larvae die. First instar larvae are very small: second instar larvae are about the size of adults, spindle shaped, and wingless. They feed actively on tender leaves and fruit, especially under the sepals of young fruit. After completing the second instar, some larvae drop to the ground to pupate; others pupate in crevices and curled leaves on the tree. The third and fourth instar thrips (prepupa and pupa) have wing pads. These stages do not feed and complete their development in debris on the ground or in the crevices of the tree. Emerged adults move actively around the tree foliage.

Citrus thrips do not develop below about 17°C (62.5°F). It has not yet been determined how many degree-days are required for development from egg to egg; however, the development takes about 1 month in cool weather, only 13 to 17 days in hot weather (see also Accumulating Degree-Days, pp. 14). The thrips can produce up to eight generations during the year if the weather is favorable. The first and second generations are usually well defined, but succeeding generations overlap (Figure 56).

Damage

The citrus thrips puncture epidermal cells, leaving scabby, grayish or silvery scars on the rind. Second instar larvae do the most damage because they feed mainly under the sepals of young fruit. As the fruit grow, the damaged rind tissue moves outward from beneath the sepals as a conspicuous ring of scarred tissue. The feeding affects the appearance but not the internal quality of fruit.

Developing fruit are most susceptible to scarring from petal fall until they are about 4 cm (1½ inch) in diameter. Heavy populations of thrips can silver the rind tissue of older fruit, but as the fruit change color, the damage generally is no longer visible. Thrips damage occurs mostly on fruit on the outside canopy, where fruit is also exposed to physical stresses such as wind scarring and sunburn.

Oviposition sites often appear during the summer, but may be a problem only on early harvested navel oranges. To expedite color development, these fruit are often gassed with ethylene, which brings out the oviposition damage more than the natural coloring process. Oviposition sites are yellowish with a brown, scabby center. They can be confused with leafhopper feeding, although leafhoppers leave uniformly yellowish or brownish spots.

On coastal lemon during the summer, other pest damage may be mistaken for thrips feeding. Rust mite or broad mite feeding causes scars that extend irregularly over the fruit; thrips feeding, however, typically results in ring scars around the stem or blossom end. Botrytis rot causes scars and ridges on the sides and shoulders of lemon fruit. Persistent wind also may cause scarring where the fruit rub against twigs or thorns.

On young leaves, thrips feeding causes thick, gray streaks on both sides of the midrib; often the leaves become distorted as they expand. Even extensive leaf damage on young trees, however, has shown no effect on their growth.

Management Guidelines

Prevention and Biological Control. A number of natural enemies attack the citrus thrips. *Euseius hibisci*, a predaceous mite that feeds on citrus red mite, shows potential as an effective natural enemy of the citrus thrips. Various spiders, the minute pirate bug, and a predaceous mite, *Anystis agilis*, also prey on thrips. Natural enemies

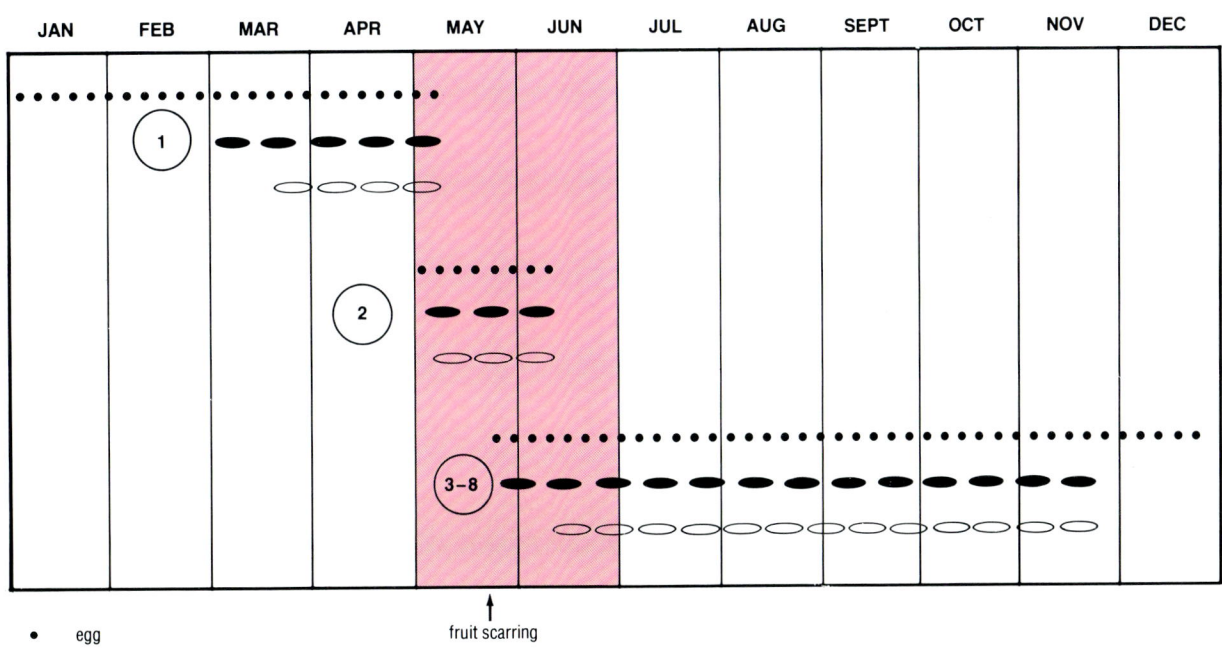

Figure 56. Seasonal development of citrus thrips on navel orange in the San Joaquin Valley. Citrus thrips produce up to 8 generations if the weather is favorable, but the most important generation is the second (2) because it scars the small fruit until they are about 4 cm (1½ inch) in diameter (red shaded area).

The first instar larva of the citrus thrips (bottom) is very small; the second instar larva (top) is about the size of the adult thrips. These instars are wingless and their antennae extend forward.

The fourth instar (pupa) of the citrus thrips has wing pads and antennae that are bent backwards.

Euseius hibisci, a predator of the citrus red mite, can be seen feeding on citrus thrips. Here the predaceous mite attacks a thrips larva.

in the ground litter feed on citrus thrips, but their impact is not known.

Orchard management practices and chemical treatments can affect thrips populations. Observations have shown that citrus thrips are less a problem in minimally treated orchards than in heavily treated orchards, where thrips populations tend to increase after treatments with certain organophosphates. Groves with a ground cover seem to have fewer thrips problems (see Ground Cover, p. 25).

Monitoring and Control. Monitoring for citrus thrips involves checking fruit samples for thrips or thrips scars. Monitoring continues as long as small, susceptible fruit are on the tree. For oranges, the monitoring period extends for 6 to 8 weeks in the spring; for coastal lemon, monitoring continues throughout the summer.

Oranges. Monitor young orange fruit for citrus thrips weekly during the critical injury period; during petal fall, monitor twice weekly. Check at least 20 outside fruit per site, following a monitoring scheme such as the one outlined under Monitoring Methods (p. 55). Monitor during midday when thrips are most active. It is best to pick the fruit and look closely under the sepals for the citrus thrips. Pay particular attention to the larvae, as they cause most of the damage. Count fruit with one or more larvae as infested and record your results in a form such as the one in Figure 57. The following guidelines have been found to be practical for navel and Valencia oranges:

• Just after petal fall, a 5% to 10% infestation may warrant a treatment.

• Once the fruit reach 1.8 cm (¾ inch) in diameter, a 20% infestation can be tolerated.

You may have to adjust these guidelines, according to the choice of insecticides available, orchard history, and predator population present.

Where possible, use the botanical, sabadilla, which provides adequate kill of moderate thrips populations while preserving natural enemies. Sabadilla acts slowly and is most effective when applied during warm, sunny weather when the thrips are most active. In the San Joaquin Valley, thrips control relies mainly on the use of synthetic organic materials, which achieve better kill but are also more destructive to beneficial species.

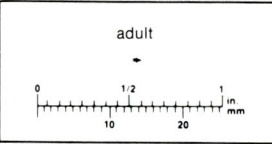

The larva of the orange tortrix is greenish to bright yellow or pale straw colored and has a golden brown head and prothoracic shield. Unlike the omnivorous leafroller, the mounds or tubercles at the base of the bristles are not chalky white.

The young larvae of the orange tortrix often feed around the fruit button.

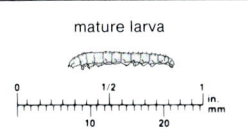

ORANGE TORTRIX
mature larva

Exochus sp. is a common parasite of the orange tortrix. The adult wasp is about 7 mm long and has a black head and body and light brown legs.

Bacillus thuringiensis, is effective against the fruittree leafroller during warm or dry weather, but is not effective against the citrus cutworm. For insecticide recommendations, see the latest University of California *Treatment Guide for California Citrus Crops*.

Orange Tortrix
Argyrotaenia citrana

The orange tortrix is sometimes a pest on Valencias and navels in southern California; it is rarely a problem on coastal lemons. It also occurs on many other plants, for example, willow, goldenrod, grapes, and strawberries. A closely related species, the garden tortrix, may also feed on citrus in southern California. Its appearance and biology is similar to that of orange tortrix.

Description and Seasonal Development

The orange tortrix occurs in overlapping generations, and larvae may be found throughout the year. In intermediate districts, the tortrix produces two or three generations, and part of the larval population aestivates during the warmest weather. The orange tortrix has adapted well to the cool coastal areas, where it may have more than three generations. Tortrix populations begin to increase in the spring, reach a peak in late summer, and decline during the winter because of adverse weather and natural enemies.

The female lays 50 to 150 eggs, overlapping them like fish scales on smooth surfaces such as stems, fruit, and the upper surface of leaves. The larvae feed inside nests spun around plant parts. Like other tortricid larvae, they wiggle and drop, suspended on a silken thread, when disturbed. They can be distinguished from other orangeworms by the color of the tubercles (see photos and key). The larvae pupate inside the nest in a dense cocoon.

Damage

The damage symptoms of orange tortrix feeding are similar to those of the fruittree leafroller. The first generation larvae roll young leaves and feed on them. The second generation larvae appear when the growth is hardening off and move preferentially to the young fruit and feed around the button. Such feeding causes only superficial scars. Later generations of the tortrix feed among clusters of ripening fruit, eating holes into the rind. They rarely feed deeper than the albedo, but the holes allow decay organisms to enter, and the fruit usually drop within 3 to 4 weeks.

Management Guidelines

Several parasites and predators attack the orange tortrix. Two parasitic wasps, *Apanteles aristolilae* and *Exochus* worm, whose adults emerge and lay eggs from late winter through the spring.

The larvae of the fruittree leafroller are green, somewhat flattened, and have a shiny black head. (For other distinguishing features, see the key in the introduction to orangeworms.) The larvae tie or roll leaves or blossoms together with silken threads and feed inside the nest. They also spin webs among mature Valencias and feed on the fruit. Larger larvae construct a new nest frequently, often daily. Mature larvae pupate inside the nests or in thin cocoons on branches or the trunk. Moths emerge about 8 to 12 days later. They mate after a few days, and the females lay their egg masses mainly on twigs 1.3 cm to 3.8 cm (½ to 1½ inch) thick in the upper part of the trees. The eggs are in a resting state until the following spring.

Damage

The larvae of the fruittree leafroller can cause substantial damage in the spring by feeding on the flush of foliage, on newly set fruit, and also on ripening Valencias, navels, or grapefruit. Early in the spring, the young larvae feed mostly on new growth flushes, often resulting in curled leaf terminals. Later on, the larvae bore into the fruit. The injury provides entry sites for secondary decay organisms. Economic loss results from fruit drop or low grades at the packinghouse. The fruittree leafroller causes damage about the same time as the citrus cutworm.

Management Guidelines

Natural enemies specific for the fruittree leafroller are not known. General predators prey on small larvae, and *Trichogramma* spp. may parasitize the eggs.

You can monitor the fruittree leafroller while you are in the orchard checking for the citrus cutworm, but you must count the two species separately. The fruittree leafroller is fairly easy to detect on the outside canopy, whereas the cutworm tends to hide among the inside leaves and fruit. Use the larval search method as outlined in the introduction to orangeworms; the sweep net shake method is not suitable for leafrollers. For assessing a population before any damage occurs and for timing a treatment most effectively, monitor the orchard for egg masses and percent egg hatch (see Monitoring Orangeworms).

Hillside slopes require special attention; egg hatching begins first in the warm, elevated areas. Watch closely those orchards that have maturing fruit on the trees. If you find two unhatched egg masses per tree during a 10-minute search in the early spring, a control action is warranted. A treatment is most effective at 25% to 50% egg hatch. If you count larvae, about 200 larvae per hour search will cause economic loss. When considering treatment during bloom, be aware of regulations protecting honeybees (see p. 36). The selective microbial insecticide,

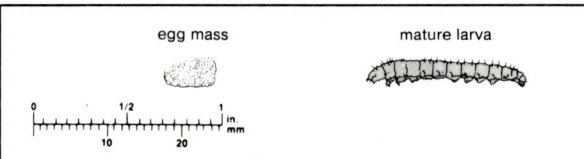

Fruittree leafroller egg masses are laid on twigs and small branches, mostly in the upper one-fourth of the tree. The lower egg mass in the photo shows the exit holes left by the emerged larvae.

Trichogramma is an egg parasite of many lepidoptera species, including the fruittree leafroller.

sp., are the two most common species. These wasps lay their eggs in tortrix larvae, and the parasites develop inside. *Apanteles* pupates in a white cocoon outside the killed larvae, whereas *Exochus* pupates inside the tortrix larva. *Exochus* emerges through a round exit hole, whereas the tortrix moth emerges through a slit at the side of the pupal case.

Monitor the orange tortrix throughout spring and summer, particularly from May through July, at 7- to 10-day intervals. Look for tortrix larvae mainly on the south and east quadrants of trees, following the larval count method outlined in the introduction to the orangeworm section. The control action threshold depends on the cultivar and the level of biological control; the threshold is about 15 larvae per hour of search for oranges and grapefruit and 30 larvae per hour of search for lemons. For pesticide recommendations, see the latest University of California *Treatment Guide for California Citrus Crops*.

On the omnivorous leafroller the tubercles or mounds, from which bristles arise along its back and sides, are characteristically chalky white. The main blood vessel along the back is visible as a faint dark stripe.

Omnivorous Leafroller
Platynota stultana

The omnivorous leafroller feeds on many different fruit trees, grapes, row crops, ornamentals, and weeds. Occasionally it causes damage on citrus in the San Joaquin Valley and in interior and intermediate districts of southern California. It generally does not occur on the coast. In the desert, the omnivorous leafroller may be a pest on nursery trees.

Description and Seasonal Development

The omnivorous leafroller produces five to six generations a year, depending on the temperature. The omnivorous leafroller typically does not build to high populations, although it may be more numerous in the late summer than in spring. Females lay their eggs overlapping like fish scales on the upper surfaces of leaves and on fruit.

The larva of the omnivorous leafroller resembles other tortricid species, particularly the orange tortrix, but has white tubercles at the base of the bristles (see photos and key). Early instars have a black head and prothoracic shield; later instars have a light brown head and prothoracic shield. The larvae roll and tie leaves together or to fruit with silken threads. When mature, they pupate inside the rolled leaves within a cocoon. They remain active throughout the winter, although their activity is much reduced.

Damage

In the spring, small larvae of the omnivorous leafroller spin webs and feed on new foliage. Later on, they tie leaves to fruit and feed under the buttons, leaving ring scarring similar to that of citrus thrips. In the summer and

During the summer and fall, the omnivorous leafroller larvae feed on new growth flushes and the peel of maturing fruit. Scarring of the peel, especially if caused by young larvae, is usually superficial; damaged leaves may have ragged edges.

The eggs of the omnivorous leafroller are disc shaped and laid in overlapping clusters.

This is a cocoon of an *Apanteles* wasp, which parasitizes many lepidopterous species, including the omnivorous leafroller.

The mature larva of the tussock moth is distinctive with its tufts of hair and its color pattern.

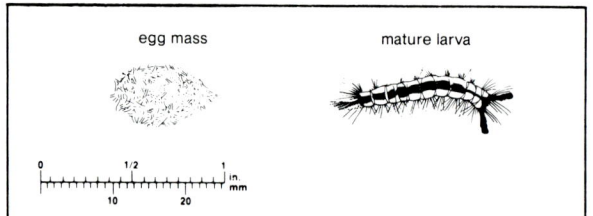

Young larvae of the tussock moth have just emerged from an egg case. They are black and have long bristles.

fall, they tie leaves to the ripening fruit and feed on the peel.

Management Guidelines

Biological Control. Several parasites attack the larva of the omnivorous leafroller. The most common are a tachinid fly and an eulopid wasp. *Erynnia tortricis* is a small, black fly, which also parasitizes other lepidopterous larvae. Tachinid females lay one or several eggs on or near the head of the larva, and the emerging maggots bore into the host, and develop and pupate inside. The larvae of the eulophid wasp, *Elachertus proteoteratis*, are gregarious, external parasites. They feed on the outside of the host larva and pupate on or near the killed larva. The parasites are most effective during midsummer. At times, *Apanteles* sp. parasitizes the larvae, and *Trichogramma* sp. may attack the eggs of the omnivorous leafroller.

Monitoring and Control. Monitor the omnivorous leafroller weekly from the spring through the fall. Look for the larvae on the south and east quadrants of trees, using the larval search method outlined in the introduction to the orangeworms section. In the spring, look for small larvae of the omnivorous leafroller under sepals when you monitor for citrus thrips. During the summer, less frequent monitoring may be sufficient, but check to see if the parasites are effective. You can tolerate a higher number of larvae in the spring, when they feed mainly on young leaves, than in the fall, when they are more likely to damage ripening fruit. No recommendations have been developed, but most materials used for other orangeworms provide adequate control of the omnivorous leafroller.

Western Tussock Moth
Orgyia vetusta

The western tussock moth is found on many orchard trees, such as apple, cherry, prune, walnut, avocado, and citrus, as well as on live oak trees, perennial lupine, and certain ornamentals. It is most common in southern California, although it occurs in a few locations in the San Joaquin Valley.

Description and Seasonal Development

The western tussock moth has one generation a year. The overwintered eggs hatch about the time the spring growth flush is expanding. The average time from egg hatch to adult is 75 days.

The larvae are distinctive: young larvae are black with long bristles; maturing larvae have numerous red and yellow spots and four median dorsal tufts. Mature larvae spin their cocoons and pupate mainly on scaffold branches and trunks, but also on other objects in the orchard, such as orchard heaters and fence posts. Adults emerge from late April through July and mate immediately after emergence. The wingless female lays 125 to 300 eggs in a single egg mass, usually on the empty pupal cell. These eggs do not hatch until the following spring.

Damage

A heavy infestation of western tussock moth may destroy all the new spring growth flush. The larvae may also eat into the newly set or young fruit; they rarely feed on mature fruit. The damage is similar to that of katydids and grasshoppers.

Management Guidelines

A dermestid egg predator, *Trogoderma sternale*, is common in certain coastal-intermediate areas in southern California, where its larvae and adults may destroy as many as 50% of the egg masses. Dermestid beetles are common from March through September, and the larvae have been observed throughout the year. A small parasitic wasp, *Dibrachys* spp., is also often seen moving about the egg masses.

Monitor the western tussock moth by looking for egg masses or larvae (see introduction to Orangeworms section). Monitor egg masses and hatching larvae to determine the population level before damage has occurred. Also, if control is needed, it is more effective on newly hatched larvae. If you find an average of one healthy egg mass per tree, economic loss can be expected from the larval feeding. Treat after about 90% of the eggs have hatched. If you monitor the larvae, about 100 larvae an hour warrants treatment. If an application is necessary during bloom, observe bee protection requirements (see page 36). For pesticide recommendations, see the latest University of California *Treatment Guide for California Citrus Crops*.

Amorbia
Amorbia essigana

Amorbia, also called the western avocado leafroller, is known as a pest of avocado, but has recently caused some damage in a number of citrus groves in southern California and the San Joaquin Valley. The pest has not been well studied on citrus, so most information on amorbia comes from studies on avocados.

The larva can be distinguished from other orangeworms by the two dark horizontal lines on each side of the

Dibrachys, a parasite of the tussock moth, is often seen moving about the egg masses.

A distinguishing feature of the amorbia larva is a black stripe on each side of the head and prothoracic shield.

Amorbia larvae tie leaves to fruit and feed on young or maturing fruit. Damaged fruit often develop decay at the feeding sites.

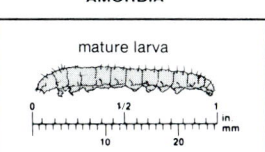

A tachinid fly has attached its eggs to the head of an amorbia larva; the emerging maggots will bore into the larva.

If disturbed, the larva of the black anise swallowtail sticks out its orange-colored scent glands and gives off a strong odor.

The parasitic wasp, *Hyposoter*, often attacks the black anise swallowtail. The cocoons of the parasite resemble small bird droppings. The shriveled skin of the host larva usually remains attached to it.

The larvae of the citrus looper are easily recognized by their looping movement.

PINK SCAVENGER CATERPILLAR

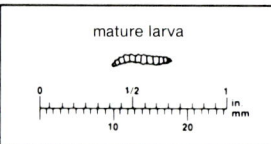

HYPOSOTER

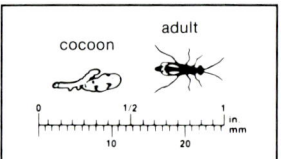

head and prothoracic shield. The female moth lays its light green eggs in a flat, overlapping egg mass of 5 to 50 eggs along the midrib on the upper surface of the leaves. Amorbia produced three overlapping generations a year in southern California and two generations in the San Joaquin Valley.

The larvae feed on new growth flushes, often rolling the leaves or tying leaves to fruit and feeding on the peel. They may also eat holes through the rind into the pulp. Damaged fruit are prone to secondary damage by decay organisms and subsequent fruit drop.

A variety of natural enemies attack the egg, larval, and pupal stages of amorbia on avocado. One of the most effective parasites of amorbia eggs is the tiny wasp, *Trichogramma platneri*. Eggs parasitized by *Trichogramma* are black. A tachinid fly, as well as several parasitic wasps, attack the larval stages. The tachinid fly, probably the same species that parasitizes the omnivorous leafroller, attaches its eggs near the head of the larva. The emerging maggots bore into the amorbia larva and develop and pupate inside. These parasites may keep amorbia populations below economically damaging levels if not destroyed by regular use of broad-spectrum insecticides.

Monitor amorbia by searching for larvae. (See introduction to Orangeworms section). Look for webbing and leaf rolls in young foliage and feeding damage on young and mature fruit on the outside canopy in the spring. The damage and threshold level are similar to those of the orange tortrix.

A pheromone trap is available to monitor the flight of the male moths, but no information on the flight pattern in citrus is currently available.

infestation will result in fruit drop or decaying fruit during storage, and treatment is rarely needed.

The black scavenger caterpillar, *Holocera iciryaceela*, occurs occasionally on coastal citrus. It is a scavenger and generally does not cause damage. The larva is easily recognized by its black color.

Black Anise Swallowtail
Papilio zelicaon

The black anise swallowtail, or California orangedog, is a native butterfly that has adapted to feeding on introduced perennial anise (sweet fennel) and citrus. On native plants, the swallowtail is single brooded, but it has multiple broods on anise and citrus. During its development, the larva changes from a mottled brown to a whitish green and bright green with yellow and black spots on each segment. Mature larvae are about 4 cm (1½ inches) long. When disturbed, all larval stages stick out orange-colored scent glands and give off a strong odor. The swallowtail feeds on tender citrus leaves, occasionally defoliating young trees but rarely causing economic damage in mature orchards. Parasites are often highly effective, especially *Hyposoter* sp. If treatment is needed, *Bacillus thuringiensis* generally provides sufficient control. Studies with sweet fennel as a trap crop have shown great promise. The black anise swallowtail prefers sweet fennel, which is also attractive to many natural enemies but not to any citrus pests.

Pink Scavenger Caterpillar
Sathrobrota rileyi

The pink scavenger caterpillar occurs sporadically in the coastal areas of San Diego and Orange counties. The larva, when fully grown, is much smaller than that of the other orangeworms. It has a light brown head, black mouthparts, a dark brown prothoracic shield, and a dark pinkish abdomen.

On orange and lemon trees, the caterpillar is mainly a scavenger, feeding on dry or decaying fruit, dead floral parts, and sooty mold. You can find the larvae mainly among fruit clusters and under the sepals. During the winter and spring months, most of the larvae stay in mummified fruit on the ground or in the tree. During the summer, the larvae may nibble on the rind of ripe Valencias, often near the stem end or on the sides of fruit in a cluster. The feeding is usually superficial and does not cause appreciable damage. Observations indicate that only a heavy

Citrus Looper
Anacamptodes fragilaria

The citrus looper is a native looper or measuring worm and occurs at low levels in most citrus-growing areas, usually together with other orangeworms. Its larvae have no prolegs in the middle of the body and therefore move in a characteristic looping or measuring fashion. The larvae consume mainly new growth flushes but also feed on blossoms and young fruit; they rarely damage mature fruit. The very young larvae typically feed on the lower leaf surface along the leaf margin. Mature larvae, which may be up to 4 cm (1½ inch) long, eat holes in the leaves or consume them entirely. The female lays about 100 pale green, spherical eggs singly on leaves. There are several generations a year. The citrus looper has many natural enemies, including *Apanteles* sp. Treatment is rarely required.

The citrus red mite has large white bristles arising from prominent, red bumps on the back and sides.

The eggs of the citrus red mite are red and have a stipe arising from the top center. With a good hand lens, you may see the fibrils extending from the stipe to the leaf surface.

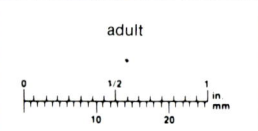

A characteristic symptom of citrus red mite feeding is the pale stippling on the upper leaf surface.

MITES

Mites are smaller than most insects; they usually have eight legs, and, unlike insects, their bodies are not segmented. A six-legged larva hatches from the mite egg; subsequent nymphal and adult stages have eight legs, except for the rust and bud mites, which have only four. Most mites are plant feeders, sucking plant juices. The plant-feeding mites can be grouped into web-spinning spider mites (twospotted and sixspotted mites) and non-webbing mites (the citrus red mite, broad mite, rust mite, and bud mite).

Water stress often aggravates mite problems. Make sure trees are well irrigated, particularly in the late summer and early fall when poor water penetration is common especially in San Joaquin Valley citrus groves. Groves on shallow soils on rolling ground and hillsides are prone to water stress because of the low water-holding capacity. Adjust your irrigation schedule according to orchard conditions.

Dust often leads to mite outbreaks. You can minimize dust by oiling orchard roads and by driving slowly and only when necessary through the grove. Where a ground cover is maintained to reduce soil erosion, dust is usually not a problem.

A number of predators attack the citrus red mite and spider mites and can keep spider mites below economically damaging levels (Table 4). Their effectiveness on citrus, however, has been limited by the frequent use of broad spectrum insecticides for citrus pests, such as citrus thrips, California red scale, and citrus cutworm.

Citrus Red Mite
Panonychus citri

The citrus red mite is one of the most important mite pests on citrus. It was first a problem only in coastal southern California; later it became important in the intermediate valleys and eventually in the San Joaquin Valley.

Description and Seasonal Development

The adult female of the citrus red mite is oval and globular; the male is considerably smaller and tapered at the rear end. Males are often found near molting female nymphs, waiting to mate. Each female lays from 20 to 50 eggs at a rate of 2 to 3 a day, depositing them on both sides of leaves. The life cycle from egg to egg may be as short as 12 days, depending on the temperature.

Populations increase in the spring, late summer, and early fall in response to new growth flushes; citrus red mites prefer to feed on young leaves. The mites will also infest fruit. On young orange and lemon trees, mites may be numerous throughout the year because repeated growth flushes provide preferred food almost continuously.

Damage

The citrus red mite extracts cell sap from leaves and fruit. On leaves, the feeding results in a pale stippling on the upper surface; the stippling is not apparent on the lower surface. In severe infestations, the stippling enlarges to dry, necrotic areas. Eventually, leaves may drop and twigs may die back.

The stippling or silvering of green fruit disappears when the fruit change color. When large populations feed on nearly mature fruit, the silvering may persist. Premature fruit drop as the result of hot, dry weather in late spring has sometimes been incorrectly attributed to mite activity. An infestation, however, can increase fruit thinning if mite feeding markedly reduces tree vigor.

The damage potential of the citrus red mite depends on tree health, water management, and weather conditions. For example, a relatively low population of mites may cause substantial damage on trees weakened by disease or lack of water. Mite damage appears mostly in late summer or fall when trees are likely to suffer from hot, dry conditions.

Management Guidelines

Prevention and Biological Control. Various predaceous mites and insects and a virus play important roles in regulating citrus red mite populations, but their effect varies with location and orchard conditions. Several management practices help preserve the predators (see page 27).

The most important natural enemy of citrus red mite is a predaceous mite, *Euseius hibisci*. This predator often remains unnoticed on the underside of leaves in the interior of the tree. It stays in the trees even if spider mites are scarce, whereas winged predators fly away. *E. hibisci* can establish its population before the citrus red mites are numerous because it has alternate food sources, such as wind-borne pollen, thrips larvae, crawlers of the California red scale, nectar, and honeydew. In the fall, however, it is slow to respond to a population increase.

E. hibisci attacks mainly the immature stages of the citrus red mite. The female predator is about the size of the female of the citrus red mite, and is drop shaped, shiny, and translucent. It lays its eggs near the midrib on the underside of leaves, in leaf depressions or in the webbing created by small spiders or psocids. The eggs are clear, oval, and about twice the size of citrus red mite eggs. Eggs develop into adults in about 8 days at optimum temperatures.

Other predators of the citrus red mite include *Stethorus*, dustywings, and green and brown lacewings. *Stethorus picipes*, a small, black lady beetle, is a general mite

Euseius hibisci, an important predator of the citrus red mite, attacks a female red mite.

The predaceous lady beetle, *Stethorus picipes*, feeds on citrus red mites. Its whitish, oval egg (right, just above leaf midrib) is much bigger than the red, globular citrus red mite eggs also shown in this picture.

A larva of *Stethorus* preys on a citrus red mite.

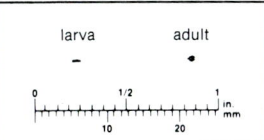

92 INSECTS, MITES AND SNAILS

Dustywings are general predators that are at times effective in reducing citrus red mite populations. The larva is conspicuous with its black and white pattern.

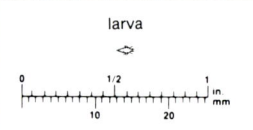

Cocoons of dustywings are usually attached to the undersides of leaves.

A virus often infects and kills citrus red mite. Infected mites move slowly, curl their legs under the body, and finally die.

predator. It does not become numerous until mite populations have reached fairly high levels. Although *Stethorus* is an important predator, it will not stay in an orchard when the citrus red mite population is low. A predaceous dustywing, *Conwentzia barretti*, is a small insect, about 3 mm (⅛ inch) long, whose wings and body are covered with a whitish powder. Its larva preys on the citrus red mite in southern California and can effectively reduce an infestation, but its occurrence is sporadic. The sixspotted thrips, *Scolothrips sexmaculatus*, which feeds mainly on the twospotted spider mite, preys on citrus red mites only when their populations are large.

Other natural control agents can have an impact on citrus red mite populations. A disease caused by a virus specific to the citrus red mite is widespread in the citrus-growing areas. Virus-infected mites show various symptoms, depending on the stage of disease development. The mites walk stiffly, curl their legs under their body, and finally die of diarrhea. In dry weather, they dry up quickly; under humid conditions, they disintegrate and leave reddish brown to black, watery spots on fruit or leaves.

Another diagnostic symptom of virus-infected mites are the internal birefringent crystals that shine in polarized light. You can see the crystals when you mount the mites on a slide and examine them under a polarizing microscope or with a specially adapted flashlight and hand lens.

The viral disease becomes epidemic only under warm, moderately humid conditions and among large mite populations. Then the virus rapidly reduces an infestation, but usually only after the red mite population has reached high levels. Careful observation of mite density, disease progress, and weather conditions are necessary to decide whether the virus can be relied upon for control or whether supplementary treatment is required.

Besides the viral disease and the predators, hot temperatures and low humidity also reduce citrus red mite populations. In the San Joaquin Valley, hot afternoons above 32°C (90°F) are common during the summer and are sufficient to suppress the mites. In southern California, hot spells of several days in late spring, summer, and fall, often accompanied by the dry Santa Ana wind, can reduce a red mite population quickly.

Monitoring Guidelines. To plan your monitoring program for the citrus red mite, survey the orchard in February or early March to see if mites are present. While walking through the orchard, scan several leaves per tree at various sites and check a few leaves for eggs and immatures with a hand lens. Start a regular monitoring program in mid-March or when mites are present. You can use the general monitoring scheme outlined under Monitoring Methods (page 55). Sample five leaves per tree in sites of four trees. Pick half- to full-sized leaves from the latest growth flush. You may want to sample more trees if you notice that the infestation is uneven, or sample more leaves per tree if the population level is approaching the threshold. Occasionally check a leaf with a hand lens to confirm your identification.

Determine the average number of adult female mites per leaf, then count the predatory mites on the underside of the leaves. The ratio of female red mites to predators will help you assess the need for treatment (see below). Note your results on a record sheet, such as the one given in Figure 61. Generally, sampling once every 2 to 3 weeks is sufficient as long as you find less than one adult female mite per leaf in the spring. If you count two adult female mites per leaf, start monitoring once a week. Continue monitoring weekly through the summer and fall if necessary.

While counting mites and predators, also record the percentage of virus-infected mites. Watch for the symptoms described under Prevention and Biological Control. At low mite densities (about one mite per leaf), you will notice hardly any virus activity. Once mite density exceeds three to four mites per leaf and the virus is present, an epidemic is likely to spread among the mites, and your weekly sampling will show a rapid increase in virus-infected and dead mites. When 40 to 60% are infected, the disease usually drastically reduces the mite population, but sampling should be continued to confirm the decline. Once the virus epidemic has spread, it overrides the effect of a predator population.

Control. Control action guidelines have been developed for the citrus red mite, but they may have to be adjusted according to time of the year, local climate, cultivar, and predator population. In the spring, threshold levels are more flexible than in the fall; predaceous mites are more abundant, and because of mild temperatures and ample moisture, trees can tolerate considerable mite feeding. During the summer, citrus red mite populations usually drop drastically except in localized areas on the coast where cool marine air allows their survival. In the fall, citrus red mites are more likely to cause damage because trees are already stressed by high temperatures and low humidity. Coastal lemons tend to have heavier red mite populations than coastal Valencias; continuous growth flushes allow a buildup of the citrus red mite while frequent pruning discourages a buildup of the predator population. In the San Joaquin Valley, predaceous mite populations have been largely destroyed by repeated use of nonselective insecticides.

Use the following guidelines with the above qualifications in mind:

• If the citrus red mite population is below 2 females/leaf and the red mite: predator ratio in the spring is about 0.2 (1 predator per 5 citrus red mites), no treatment is recommended.

• If the red mite population is above 2 females/leaf in the spring, the ratio has to be higher (0.5 or 1 predator per 2 red mites) for sufficient control. If the number of females/leaf exceeds 3 to 4 and the ratio does not rise above 1 within 10 to 14 days, a treatment may be warranted, depending on orchard conditions.

ORCHARD _____ BLOCK _____ CULTIVAR _____

COMMENTS _____

DATE	CRM (females/leaf)	predator (per leaf)	ratio (females/predat.)	virus (% mites infest.)	other mites (per leaf)	Remarks

Figure 61. Field record sheet for citrus red mite (CRM). You can use the format in Figure 57 and substitute the headings. (See text for monitoring details.)

Twospotted mites vary in appearance; the dark spots on the side of the body may hardly be visible just after molting or in immature mites.

Feeding by twospotted mites results in brown, necrotic areas between major leaf veins.

A fine webbing among fruit or leaf clusters is typical of a twospotted mite infestation.

- If the red mite population reaches 3 to 4 females/leaf in July/August, no treatment is generally required because hot weather drastically reduces the population about this time.
- If the red mite population exceeds 2 to 3 females/leaf in the fall, treatment may be warranted, depending on tree stress.

The proper treatment time and choice of materials for citrus red mite control differ with the growing region. In coastal and intermediate areas of southern California, a treatment in the spring is rarely recommended. An application, if necessary, is usually made between August and October. Low-volume oil spray is effective and allows the survival of predaceous mites. Treat navel oranges no later than September because oil has adverse effects on the maturing fruit after this time. An oil spray during this period also reduces the populations of bud mites and immature stages of the California red scale.

In the San Joaquin Valley, the citrus red mite may be a problem in the spring and sometimes the fall. Selective acaricides (see Table 5), when needed, will reduce mite populations and preserve natural enemies. For pesticide recommendations, see the latest University of California *Treatment Guide for California Citrus Crops*.

Twospotted Mite
Tetranychus urticae

Although mainly a pest of deciduous fruit and nut trees as well as many ornamentals and vegetable crops, the twospotted spider mite has become an occasional pest on citrus, particularly in the San Joaquin Valley. Its damage potential varies from year to year and seems to be related to other stress factors, such as water stress and heat.

Description and Seasonal Development

All stages of the twospotted mite overwinter in protected places on the tree, such as the navel of navel oranges, under the button, and where several fruit touch. If the weather is mild, the mites will continue to feed and reproduce during the winter. In the late spring, mite activity increases; mites are most numerous during the summer. Colonies develop in pockets on the underside of leaves and, when heavy populations build up, also develop on the upper sides of leaves and on fruit. They cover leaves and fruit with a conspicuous webbing. On mature fruit, webbing is often heavy and may completely cover several grouped fruits.

The eggs are spherical and translucent when first laid, becoming opaque before hatching. They are scattered throughout the webbing. Immature mites molt three times before becoming adults. If temperature and food supply are favorable, a generation can be completed in 7 days.

Sixspotted thrips have three dark spots on each forewing. They are effective predators of spider mites.

Damage

A light infestation of twospotted mites results in yellow or brown spots between leaf veins. Clusters of dried, brown leaves and profuse webbing among leaf and fruit clusters indicate a heavy infestation. If stress from a mite infestation is compounded by lack of water or other factors, leaves and fruit may drop.

Management Guidelines

Prevention and Biological Control. Cultural practices, such as adequate irrigation, help reduce the impact of spider mite feeding. By choosing selective chemicals when treating other pests and by reducing dust in and around the orchard, you can preserve the natural enemies of spider mites.

A number of predators can provide substantial control of the twospotted mite. Sixspotted thrips can reduce populations of the twospotted and citrus red mites, particularly in the winter and early spring. Treatments for citrus thrips in late spring, however, usually kill these predators. The spider mite destroyer, *Stethorus picipes*, is a small lady beetle that feeds on the twospotted mite as well as on the citrus red mite, but does not become numerous until mite populations are heavy. The minute pirate bug also feeds on spider mites. The western predatory mite, *Typhlodromus occidentalis*, is an important predator of web-spinning spider mites in deciduous orchards. It has recently been found to attack twospotted mites in a few citrus groves in the San Joaquin Valley. *Euseius hibisci*, a predaceous mite that attacks the citrus red mite, generally does not enter the heavily webbed colonies of the twospotted mites or other spider mites.

Monitoring and Control. In the San Joaquin Valley, check for twospotted mites when you monitor the citrus red mite in late winter and early spring. Continue monitoring the twospotted mite occasionally during the summer. Monitor more closely in late summer and fall. Examine the leaves in areas under water stress and at the edges of the orchard where mites may be more numerous because of dust. Also look for brown clusters of mite-damaged leaves throughout the orchard. Colonies usually begin on the lower branches in the inner canopy, so look there first. In the summer and fall, high populations of mites may require treatment. Threshold levels, however, have not yet been established. For pesticide control recommendations, see the latest University of California *Treatment Guide for California Citrus Crops*.

Broad Mite
Polyphagotarsonemus latus

The broad mite occurs only in coastal growing areas, where in recent years it has become a problem in many lemon orchards. This mite is so small that you need a hand lens to see it. The adults are fairly broad and have eight legs. The males have much longer hind legs than the females. Broad mites are often found in depressions on the fruit, where the females lay their eggs. You rarely see broad mites in the spring, when predaceous mites keep the populations low. Native predaceous mites feed on the broad mite, but a recently established species, *Amblyseius stipulatus*, may be an even better predator. The broad mite is most numerous from late July through early October.

A related mite, *Tarsonemus cryptocephalus*, should not be confused with the broad mite. This species is not a pest; it feeds on honeydew, sooty mold, dead scales, and other decaying organic matter. Often these mites accumulate under fruit buttons or empty scale covers. Besides their different habitat on the tree, the two mite species also differ in their coloration: *Tarsonemus* is dark amber, and the adult female has a white spot on the posterior; the broad mite is yellowish or light amber, and the adult female has a white stripe on its back. Both mite species are transparent just after molting.

Broad mites feed on fruit and leaves. Their preferred feeding site is the shady side of young fruit—up to about 2.5 cm (1 inch) in diameter—inside the canopy. When populations become dense, the mites move to outside fruit. The feeding results in scarred tissue that cracks as the

The broad mite is very small, fairly broad, and has eight legs and a white stripe down its back. The eggs (top right) have prominent white mounds and are laid in depressions of the fruit.

Broad mites usually feed on the inner or shady side of fruit. As the fruit grow, the scarred tissue breaks up into a net pattern with intermittent new tissue.

fruit grows, leaving a characteristic pattern of scars and intermittent new tissue. Although most damage occurs on the fruit, broad mites may also feed on young, expanding leaves, causing them to curl. The curling and crinkling of leaves may at times resemble bud mite damage, although bud mite feeding normally results in a rosettelike growth.

Monitor the broad mite from July through September. No detailed management guidelines have yet been developed. If you find a medium-size, increasing population, a treatment is usually indicated. Choose materials least disruptive to an IPM program. Two commonly used acaricides—dicofol (Kelthane®) and sulfur—are very toxic to predaceous mites, but oxythioquinox (Morestan®) and cyhexatin (Plictran®) are less toxic and provide adequate control. For pesticide recommendations, see the latest University of California *Treatment Guide for California Citrus Crops*.

Citrus Bud Mite
Eriophyes sheldoni

The citrus bud mite is a pest mainly on coastal lemon. This mite is very small, elongated, somewhat tapered at the posterior end, and has four legs at the front end near the mouth. The female lays about 50 eggs mostly in the bud scales of last year's growth. The emerging young mites feed inside the bud, killing it or causing a stunted, rosettelike growth of leaves or oddly distorted fruit. At times, the stunting and crinkling of leaves may be confused with broad mite damage, but broad mite feeding does not cause the typical rosetting of leaves. A bud mite population can increase rapidly in infested buds or under fruit sepals during the summer. After a peak in late August or September, it gradually declines. No effective natural enemies of the bud mite are presently known.

To detect bud mites before damage occurs, check the buds on last year's growth (15 to 20 cm or 6 to 8 inches inside the canopy) in late summer. Collect five buds each from about 20 randomly chosen green twigs throughout the orchard. Examine them under a microscope or with a 20x hand lens for signs of bud mite activity, i.e., necrotic tissue, cast skins, dead or live mites. No threshold has been established; however, observations suggest that if 10% of the sampled buds are infested with live mites, economic damage through loss of fruit buds and distorted fruit is likely. For pesticide recommendations, see the latest University of California *Treatment Guide for California Citrus Crops*.

Citrus Rust Mite
Phyllocoptruta oleivora

The citrus rust mite is a sporadic pest in coastal citrus plantings. It is called silver mite on lemon and rust mite on oranges. It is about the same size as the bud mite, but it is deeper yellow and wedge shaped. A generation may be completed in 1 to 2 weeks in the summer; development slows or stops during the winter, depending on the temperature.

The rust mite feeds on the outside, exposed surface of fruit 1.3 cm (½ inch) or larger. The feeding destroys the rind cells, and the surface becomes silvery on lemon, rust brown on mature oranges, or black on green oranges. Rust mite damage is similar to broad mite damage, except that somewhat larger fruit is affected, and the scarring is generally more uniform than that of the broad mite. Most rust

Bud mites feed and lay their eggs inside developing buds. These mites are tiny and elongated and have four legs near the mouth.

Bud mite damage may result in oddly misshapen fruit, such as this lemon.

The rust mite is tiny and wedge shaped and has four legs at the front end.

Buds damaged by bud mite feeding develop into distorted blossoms or leaves.

On lemon, rust mite feeding results in more or less uniform silvering of the rind tissue.

On mature oranges, rust mite damage causes russeting of the rind.

The adult of the citrus flat mite is flat and has variable coloration.

mite damage occurs from late spring to late summer, although damage is possible throughout the year in groves not treated for other pests. No effective natural enemies are known, although general mite predators at times feed on broad mites.

Monitor the citrus rust mite from early spring through summer. On orange trees, look for rust mites on young foliage in early spring; by late spring, most of the population will be on the fruit. On lemon, rust mites are mostly on the fruit throughout the season. Check outside fruit for scarred rind tissue. Once scarring appears, however, the mites have usually moved onto fresh, undamaged fruit; therefore also check neighboring green, small to mid-sized fruit for the presence of rust mites. A 10x to 14x hand lens is necessary to identify these minute mites. They usually feed in protected places, such as the stylar end. When populations are high, the mites move over the entire fruit.

Once you find one or more infested fruit and if rust mites were a problem the previous year, watch the orchard closely. Threshold levels depend on last year's rust mite problems and current market conditions. If the population increases quickly or if scarring appears, a treatment is generally required. In some cases, the infestation is localized and a spot treatment may be sufficient for control. For pesticide recommendations see the latest University of California *Treatment Guide for California Citrus Crops*.

Other Mites in Citrus

Flat Mite   *Brevipalpus lewisi*

The flat mite is a minor pest of citrus in the desert regions and interior valleys. It occurs in the San Joaquin Valley but is rarely seen on citrus because of treatments for other mites. The adult is smaller than the citrus red mite, is flat, and varies in color. The flat mite is usually a secondary invader, feeding on rind tissue damaged by leafhopper feeding or thrips oviposition. The mite feeding results in a scabbing of the leafhopper injury, which would otherwise disappear as the fruit change color. The flat mite is fairly heat tolerant, so populations persist during the hot summer. For pesticide recommendations, see the latest University of California *Treatment Guide for California Citrus Crops*.

Yuma Spider Mite   *Eotetranychus yumensis*

The Yuma spider mite occurs on grapefruit and lemon in the Coachella and Imperial valleys. This species is similar in shape to the citrus red mite but is pale straw to dark pink colored and produces substantial webbing. It feeds and lays its peach-colored eggs under the heavy webbing on the underside of leaves. Yuma spider mites are most numerous in winter and late spring. Although they may cause some leaf drop, damage is usually not severe enough to warrant treatment.

Sixspotted Mite   *Eotetranychus sexmaculatus*

The sixspotted mite is a minor pest on citrus in some coastal growing areas. This mite is somewhat smaller than the twospotted mite, is lemon yellow, and usually has three pairs of black spots. A generation is completed in 3 to 4 weeks, and populations are usually heaviest in the spring and early summer.

Sixspotted mites feed along the midrib or larger veins on the underside of citrus leaves. They form small colonies and cover themselves with protective webbing. A depression develops where a colony has settled and becomes apparent as a slight bulge on the upper leaf surface. The infested area may turn pale to yellow, and the leaves often become distorted. Leaf drop may occur with few mites present. Predaceous mites and sixspotted thrips usually keep sixspotted mites under control. It is a problem in some areas, however, particularly in those protected from the Santa Ana winds.

Lewis Spider Mite   *Eotetranychus lewisi*

The Lewis spider mite is a rare pest of citrus in the desert valleys of southern California. The females are about the size of the twospotted mite, vary in color, and develop a number of black spots along the lateral margins. The eggs are colorless to straw colored and have a slender stalk. The mites produce extensive webbing.

The Lewis spider mite feeds only on the fruit, mainly in depressions at the stylar end of maturing fruit. It causes a silvering of lemons or russeting of oranges. Usually the damage is not severe enough to warrant treatment. Harvesting removes most of the infestation.

Ants

As a group, ants are important natural enemies of many insect pests. In citrus, however, the detrimental effects of honeydew-seeking species outweigh their benefits for pest management. The most prevalent species is the Argentine ant, *Iridomyrmex humilis*, which has so far only been a problem in southern California where its presence hampers biological control. The ants feed on the honeydew excreted by soft scales, mealybugs, cottonycushion scales, whiteflies, and aphids. By protecting their favorite food source from natural enemies, they interrupt the natural control of these pests. For example, the ants can virtually eliminate the parasite, *Metaphycus luteolus*, allowing the brown soft scale to build up rapidly on trees or parts of trees bearing a high ant population. Similarly, the most effective parasite of the black scale, *Metaphycus helvolus*, is usually scarce on trees heavily infested with ants. Populations of woolly whitefly and the citrus mealybug also increase in the presence of ants. Ants generally do not interfere with the biological control of the cottonycushion scale and the citrophilus mealybug.

Ants may also affect certain insect pests not producing honeydew. Red scale populations build up in the presence of a heavy ant population; the ants inhibit the activities of *Aphytis* parasites and several predaceous coccinellid beetles. Ants interfere with other natural enemies, such as the larvae of syrphid flies, lacewings, and dustywings.

Two other ant species are at times a problem in citrus groves. The native gray ant, *Formica cinerea*, may return to orchards where the Argentine ant has been controlled. It also protects honeydew-producing insects against natural enemies. The southern fire ant, *Solenopsis xyloni*, does not protect pests but feeds directly on tender twigs, bark, and leaves of small trees, sometimes girdling the trees. It is attracted to the gum droplets, which often ooze from the bark of stressed trees.

Description and Seasonal Development

The Argentine ant is a small species and is uniformly deep brown. Most often seen are the workers, which travel in characteristic trails on the trees or the ground. They build their nests under the soil surface. From fall through spring, they move their nests to sunny locations; with the onset of hot weather, they build the nests under trees. The size of the colony varies from a dozen to many thousands, and the number of queens from one to many hundreds. The ants usually aggregate in large colonies in the winter and break up into small colonies during the summer. After mating in the spring, the queens lay large numbers of eggs; thus ant numbers increase enormously in midsummer and early fall.

The southern fire ant is light reddish brown with a black abdomen. These ants build nests of loose mounds

The Argentine ant feeds on honeydew excreted by soft scales, whiteflies, mealybugs, and aphids. It disrupts biological control of these pests by protecting them from attack by parasites and predators. Note the swollen, almost translucent abdomen, which identifies the species as a honeydew collector.

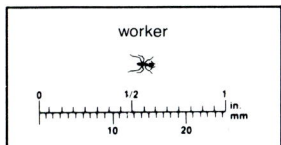

or numerous scattered craters near the base of trees. Fire ants aggregate in smaller numbers than do Argentine ants.

The native gray ant is considerably larger than the other two species and gray in color. It moves fast in irregular patterns and builds its nest in top soil or under a rock or debris.

Management Guidelines

No effective natural enemies of the Argentine ant are known; usually only competition with native ant species will limit their population.

Monitor the orchard in spring when honeydew-producing insects, such as aphids, appear. Check the trees for ants and see if they have swollen, almost translucent abdomens; this identifies them as a honeydew-collecting species. For the most effective and most economical ant control, treat when populations are low and place the pesticide near nests or on trails. Broadcasting the material is expensive, does not reach many nests, kills ant species that provide competition, and interferes with the biological control of the brown garden snail. Timing of control measures is also important: treat in the spring when most nests are still outside the drip line, usually on the south and southwest side of trees. Preliminary field tests show that placing a pesticide on trails (drip lines are often used as trails) or putting poison bait stations under trees can keep ants at low population levels with small amounts of pesticides. For pesticide recommendations, see the latest

Katydid nymphs are wingless and have long antennae with black and white bands.

Feeding by katydids causes roundish scars in the rind tissue.

Mealybugs feed in dense colonies, apparent as white, sticky clusters among leaves or fruit.

University of California *Treatment Guide for California Citrus Crops*.

If the tree skirts are pruned up, a barrier of tanglefoot around the trunks will keep the ants out of the trees but must be renewed regularly.

To prevent damage by the southern fire ant, plant trees carefully and irrigate them well; healthy, vigorous trees are less likely to exude gum, which attracts these ants. Bordeaux whitewash helps prevent gumming.

Control the gray ant in the same way as the Argentine ant.

Katydids

Katydids are sporadic pests on citrus. Of the two species feeding on citrus, only the fork-tailed katydid (*Scudderia furcata*) causes economic damage. The angular-winged katydid (*Microcentum retinerve*) is less abundant and feeds only on leaves.

Description and Seasonal Development

Katydids resemble grasshoppers but have long antennae. The nymphs are wingless and have black and white banded antennae. Katydids produce one generation a year. The females of the forktailed katydid lay their eggs into the edges of their chewing damage. Nymphs appear in April and May and require 2 to 3 months to mature. Katydids normally spend their entire life cycle on citrus trees.

The angular-winged katydid is larger, has broader wings, and has a humpback appearance, both in the nymphal and adult stages. The females lay their eggs in two overlapping rows on twigs and leaves.

Damage

Fork-tailed katydids cause most damage at petal fall, when they feed on the young fruit. The feeding sites become covered with grayish scar tissue and the expanding fruit becomes misshapen. Katydids do not eat the whole fruit; they take a bite and move on to another feeding site on the inside of fruit. Thus even a few katydids can damage a large quantity of fruit in a short time. They also eat holes in leaves and maturing fruit. The holes resemble citrus cutworm damage.

Management Guidelines

In the San Joaquin Valley, broad-spectrum insecticides used for other pests generally suppress katydids, but flare-ups do occur. In southern California, katydids are rarely a problem; in orchards with good biological control, parasites often attack the eggs of katydids.

You can detect an infestation in the fall by looking for eggs laid inside the edges of leaves. When monitoring the grove in the spring for orangeworms, also check fruit for katydid damage. At that time, you can see katydid nymphs on the outside of the canopy, but they jump readily when disturbed. If you find any katydids in an orchard with previous katydid damage, especially at petal fall, treatment is recommended. For insecticide recommendations, see the latest University of California *Treatment Guide for California Citrus Crops*.

Mealybugs

Mealybugs occur on a wide variety of fruit trees, field crops, and ornamentals, such as mulberry and catalpa trees. They can be pests of citrus in certain coastal areas, but natural enemies generally keep populations below damaging levels.

Several species occur in California, including the citrus mealybug (*Planococcus citri*), citrophilus mealybug (*Pseudococcus fragilis*), long-tailed, mealybug (*Pseudococcus longispinus*), and Comstock mealybug (*Pseudococcus comstocki*). The citrus mealybug is the most common species. The Comstock mealybug was first discovered in the San Joaquin Valley in 1967, but introduced parasites now keep this species at a low population level on the many ornamentals it attacks. The Comstock mealybug does not establish itself on navels or tangerines but can infest lemon; therefore quarantine restrictions still exist for the movement of lemons from the San Joaquin Valley into southern California. The other mealybug species are kept at low population levels by effective parasites.

Description and Seasonal Development

Mealybugs are soft, oval, flat, distinctly segmented, and covered with a white, mealy wax, which extends into spines along the body margin and the posterior end. The species differ mainly in the thickness and length of the waxy spines. The citrus mealybug has a yellow-orange body visible through the powdery wax. The filaments around the margins are not appreciably longer at the posterior end. The Comstock mealybug has a thicker wax cover and two spines at the posterior end about one quarter the length of the body. Adult males of mealybugs are small, two-winged insects with two long filaments at the posterior end.

The females lay several hundred eggs within 10 to 20 days in cottony egg sacs attached to leaves, fruit, or twigs. The newly hatched nymphs are light yellow and free of wax but soon start to excrete a waxy cover. Mealybugs produce two to three overlapping generations a year. They pass the winter mainly in the egg stage, although all other stages may be seen on the trees.

The citrus mealybug has a yellow orange body visible through the powdery wax. The filaments around the body margins are not appreciably longer at the posterior end than on the sides.

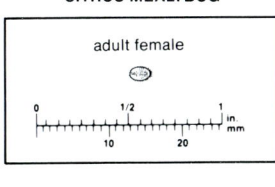

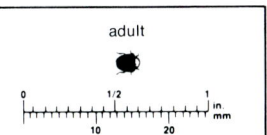

The Comstock mealybug has a thicker wax cover than the citrus mealybug, and the two spines at the posterior end are about one quarter the length of the body.

The *Cryptolaemus* adult (left) and its white larva (right) feed voraciously on all mealybugs.

The woolly whitefly got its name from the curly, waxy filaments that cover the third and fourth instar larvae. The colony is covered with droplets of excreted honeydew.

The dark color of these woolly whitefly pupae indicates that they are parasitized by *Amitus spiniferus*. The light brown body on the upper right is a nonparasitized third instar of the woolly whitefly.

Amitus spiniferus lays an egg into a young woolly whitefly.

Damage

Mealybugs extract plant sap, reducing tree vigor. If a cluster of mealybugs feeds along a fruit stem, it often results in fruit drop. The excreted honeydew provides a medium for the growth of sooty mold. Damage is most severe in spring and fall.

Management Guidelines

Natural enemies are effective in keeping mealybug populations below economically damaging levels, and treatment is rarely required. Parasites, in particular, provide good control of the citrophilus, long-tailed, and Comstock mealybugs in minimally treated orchards (Table 5). Native predators include lady beetles, lacewings, and syrphid flies. The most common introduced predator of the citrus mealybug is the mealybug destroyer, *Cryptolaemus montrouzieri*. Its larva closely resembles a mealybug but is about twice as large as an adult citrus mealybug female. It also has longer lateral waxy filaments and moves faster. Both the adult beetle and the larva feed voraciously on all stages of mealybugs. *Cryptolaemus* does not survive the winter well, and extensive annual recolonization programs were carried out for many years. The beetle is still being raised in some insectaries in Ventura County and released to supplement natural control during sporadic outbreaks of mealybugs. If a heavy population of mealybugs must be reduced quickly, parathion can be used but should be followed by a release of *Cryptolaemus* after about 30 days to reestablish natural control. Malathion should be avoided because it is generally longer lasting and more disruptive to natural control than parathion.

Synthetic pheromones for the citrus and Comstock mealybugs are available to monitor incipient populations.

Whiteflies

In California citrus, the two whitefly species of economic importance occur mainly in the coastal-intermediate areas: the woolly whitefly, *Aleurothrixus floccocus*, and the bayberry whitefly, *Parabemisia myricae*. The woolly whitefly was first discovered in San Diego County in 1966 and has since spread to other southern California counties. The bayberry whitefly is a more recent invader; it was detected in 1978 and is now present in most southern California counties. A previously important species, the citrus whitefly (*Dialeurodes citri*), is today only occasionally of concern. All whiteflies on citrus, except the bayberry whitefly, are now under partial to complete biological control. The mulberry whitefly is not a pest on citrus but is occasionally seen on southern California citrus. The greenhouse whitefly occurs frequently in orchards in the San Joaquin Valley. It is usually not a pest but produces honeydew, on which sooty mold grows.

Description and Seasonal Development

Whiteflies derive their name from the mealy white wax covering their wings. Whitefly adults are similar in appearance; the pupal stage (fourth instar) is more distinctive. The woolly whitefly pupa is covered with curly, waxy filaments, whereas the pupa of the bayberry whitefly has a clear wax fringe around its body margin. The pupa of the citrus whitefly lacks the waxy filaments and the marginal fringe but has a distinctive Y-shape on its back. The pattern of egg laying is also useful for field identification. The woolly whitefly usually lays its eggs in circles or partial circles on the lower surface of full-sized leaves. The bayberry whitefly lays its eggs randomly on both sides of young leaves, particularly on leaf margins, and also on young fruit when populations are high. The eggs of the citrus whitefly are attached randomly on the underside of full-sized leaves.

All whitefly species have similar life cycles. The first instar larvae are initially mobile (crawlers); they soon lose their legs and settle to feed. The following three larval stages are also sessile. The larvae are generally flattened and oval, resembling certain soft scales. The winged adults emerge through T-shaped splits in the pupal skin, whereas the parasites of whiteflies emerge through oval to round exit holes. Whiteflies have many generations a year and are most numerous from July to November. The bayberry whitefly, in particular, can build to high populations quickly because it can reproduce females without mating. Lemons are most heavily infested because they provide a continuous food source of new foliage.

Damage

Whiteflies suck cell sap from leaves, which wilt and drop when populations are large. The honeydew excreted by the larvae collects dust and supports the growth of sooty mold. Large infestations can almost blacken entire trees, reducing photosynthetic activity and causing defoliation. The honeydew also attracts ants, which interfere with the biological control of the whiteflies and other pests. Whiteflies usually do not damage fruit directly.

Management Guidelines

A number of parasites that attack the immature stages have been introduced for whitefly control (Table 4). The natural enemies of the citrus and woolly whitefly provide complete control when undisturbed by ants, dust, or insecticide treatment. Efforts to bring the bayberry whitefly under biological control are encouraging. Two parasites introduced from Japan are becoming established, and recently an *Eretmocerus* species of unknown origin was discovered attacking this new whitefly species.

Monitor whiteflies from April through December. The bayberry whitefly may flare up on new growth flushes in the spring and fall or after pruning, and requires the

Cales noacki parasitizes early instars of the woolly whitefly.

The female woolly whitefly lays its eggs in circles or partial circles on fully grown leaves.

The adult bayberry whitefly has pearl-colored wings and widely spaced antennae. The female lays its eggs on young leaves, particularly on the margins.

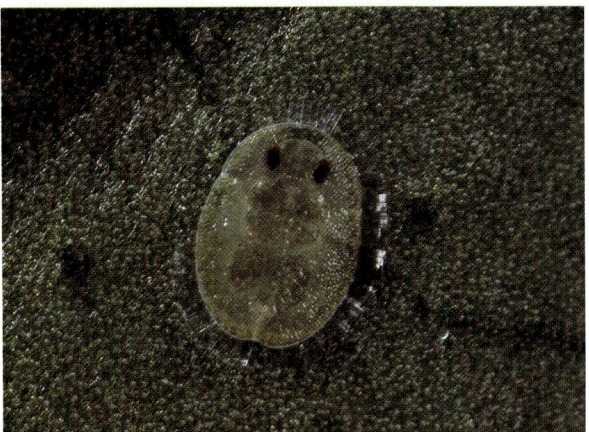
The pupa of the bayberry whitefly has a clear, wax fringe around the body margin.

A native species of *Eretmocerus* parasitizes the bayberry whitefly. It has distinctive clublike antennae.

Empty bayberry whitefly pupal skins with a round exit hole remain after emergence of *Encarsia* parasites. One skin contains the dark cast skin and meconia from the parasite. Pupal skins from which a whitefly adult emerged are clear with a T-shaped opening.

The pupa of the citrus whitefly has no waxy fringe or filaments but has a distinct Y-shaped marking.

The mulberry whitefly is sometimes seen on citrus. Its pupa is black with a white fringe.

most careful attention. Because of the high reproductive potential, short generation time, and initial lack of effective natural enemies, bayberry whitefly populations have been explosive at times.

A combination of management strategies offers the best promise for successful population suppression. Parasite releases, ant and dust control, and cultural manipulations help establish and maintain populations of the new natural enemies. Selective pruning, such as alternate row pruning, provides refuge for the parasites. Chemical treatment has not been effective; studies have shown only temporary suppression, often with a resurgence of the pest. Check with your farm advisor for the latest developments in managing the bayberry whitefly.

Other Insect Pests

Potato Leafhopper *Empoasca fabae*

The potato leafhopper is a potential pest of citrus in some areas, especially in groves near tomato fields, cotton fields, or pastures in the San Joaquin Valley. It is a green, slender insect with bristlelike antennae and rows of spines along its hind legs. It breeds in large numbers on wild plants and field crops. During the late summer and fall, the leafhoppers may migrate to citrus groves to spend the winter in the shelter of the trees.

The potato leafhopper feeds on fruit by puncturing rind cells, causing yellowish to light brown, roundish scars on fruit. The scars are particularly apparent on green fruit and resemble thrips oviposition sites except they are more clustered and do not have darkened centers.

Leafhoppers are not a problem every year. Use a yellow, sticky card, such as the one used for the California red scale or traps, to determine if leafhoppers are present. If you apply a Bordeaux spray in the fall against brown rot and Septoria, you may want to add some additional hydrated lime to repel leafhoppers. Where leafhoppers are a persistent problem, you may have to apply a separate lime treatment. For pesticide recommendations, see the latest University of California *Treatment Guide for California Citrus Crops*.

Aphids

Aphids are generally not a problem on citrus except where there are young trees. The most common species on citrus, particularly in the coastal and intermediate districts of southern California, is the spirea aphid, *Aphis citricola*. Its feeding on young growth flushes causes the leaves to curl. The cotton or melon aphid, *Aphis gossypii*, occasionally feeds on citrus and can transmit the tristeza virus. The two species can be distinguished by their color; the spirea aphid is green, whereas the cotton aphid is dark gray or dull black.

A number of predators, parasites, and fungal diseases usually keep an aphid population below damaging levels. A moderate aphid population (about 40% of growth flushes infested) can be considered beneficial on mature trees because the aphids and their honeydew provide a good food source for many natural enemies of other pests early in the season when other hosts are not available.

On newly established trees, feeding by aphids can retard and stunt new growth. Inspect young orchards frequently; if natural enemies are not controlling an aphid population, an outside coverage spray is warranted. For insecticide recommendations see the latest University of California *Treatment Guide for California Citrus Crops*. Treatment of the cotton aphid for the purpose of preventing the transmission of the tristeza virus has not been effective.

The potato leafhopper occasionally causes damage on citrus. It is a slender, green insect with a row of spines along its hind legs.

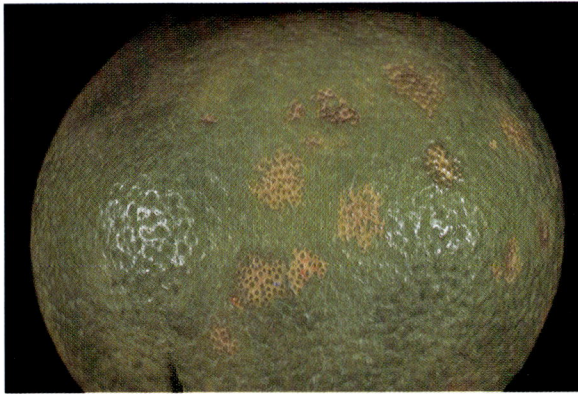

Leafhopper feeding sites are apparent as brownish scars, particularly on green fruit.

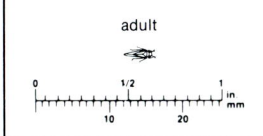
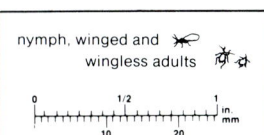

The green color of the spirea aphid, the most common aphid species on citrus, blends well with the color of the young citrus leaves on which it feeds. The winged form has a dark brown to black thorax.

Aphids feed mainly on the underside of young leaves, causing them to curl toward the stem.

Several parasites attack aphids. Parasitized aphids change their color and become sluggish. Often you can see mummified, light brown aphid bodies with a round exit hole cut by the emerging adult parasite. Here a *Lysiphlebus* sp. has just emerged from an aphid mummy.

The larvae of the syrphid fly are effective predators of aphids. Here a larva feeds on spirea aphids.

This shriveled aphid has been killed by a fungal disease that is common under warm, humid conditions.

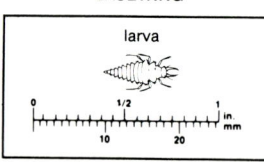

The larvae of lacewings feed on aphids, other small insects, and insect eggs.

Fuller Rose Beetle *Pantomorus cervinus*

The Fuller rose beetle occasionally feeds on citrus foliage and causes localized damage. The beetle produces a brood in the spring and in the fall, and each requires 12 months to complete its life cycle. The larvae feed on the roots of citrus trees or of weeds in and around the orchard. The adults feed along the margins of citrus leaves, resulting in a characteristic ragged appearance. The damage is mainly seen on the lower part of the trees. On large trees, feeding by the Fuller rose beetle can be ignored, but on small trees the loss of leaf surface may be significant enough to reduce photosynthetic activity and impede growth. The most severe damage, however, occurs when the beetles feed on buds, thus destroying entire growth flushes. For insecticide recommendations, see the latest University of California *Treatment Guide for California Citrus Crops*.

Grasshoppers

Various species of grasshoppers may feed on citrus, mainly in orchards bordering rangelands. Large numbers

INSECTS, MITES AND SNAILS

The adult Fuller rose beetle occasionally causes damage on young citrus trees.

The Fuller rose beetle feeds around the edges of leaves, resulting in a typical ragged appearance.

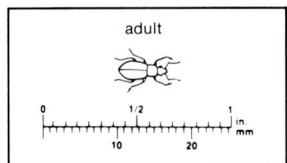

The Mediterranean fruit fly has a yellow and black body and banded wings. It is a pest of deciduous and subtropical fruit in Mediterranean countries, east and southwest Africa, western Australia, and many islands in the Atlantic and Pacific oceans.

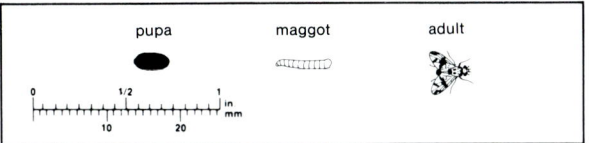

The oriental fruit fly occurs in many countries of the Far East and on the Hawaiian islands.

of grasshoppers usually hatch during April and feed on grasses. Once the grasslands dry up, they migrate to green foliage, such as citrus. They feed on the citrus fruit, leaves, and green twigs. Fruit damage is similar to that caused by katydids or orangeworms. On young trees, the loss of foliage and green wood can retard growth. To avoid damage, you can place carbaryl* bait along the borders of the orchard before the grasshoppers migrate into citrus trees or spray infested trees. A permit from the county agricultural commissioner is required for purchase and use of carbaryl.

Fruit Flies

Fruit flies, such as the Mediterranean fruit fly (*Ceratitis capitata*), oriental fruit fly (*Dacus dorsalis*), and

The brown garden snail eats deep holes into ripening fruit.

The decollate snail can be an effective predator of the brown garden snail.

Mexican fruit fly (*Anastrepha ludens*) currently do not occur in California. Inadvertant importation from areas where these fruit flies are a problem, such as Mexico or Hawaii, is a constant threat. Fruit flies attack most deciduous and subtropical fruit, including citrus. The females lay eggs in the fruit rind, and the maggots develop inside the pulp, destroying the fruit tissue.

Brown Garden Snail
Helix aspersa

The brown garden snail is an introduced European species that thrives in cultivated areas, such as gardens and orchards. Over the years, the snail has become a particular problem in citrus orchards, where changes to no-till weed control and sprinkler and drip irrigation have created an ideal environment for snail development. The brown garden snail can cause extensive damage by feeding on ripening fruit.

Description and Seasonal Development

The brown garden snail is rarely confused with other native snails because it is usually the only species present in any numbers in citrus orchards. It is about 2.5 cm (1 inch) in diameter at maturity and has a distinct color pattern. The brown garden snail is most active during the night and early morning and during cool, damp periods. In southern California, particularly in coastal areas, young snails are active throughout the year. The more mature snails hibernate in the top soil during a cold winter, but may remain active if the winter is warm. During hot, dry periods the snails seal themselves off with a parchmentlike membrane and remain on the tree or in the soil. The snails are bisexual, and all snails of reproductive age lay eggs up to six times during a season, depending on local climate and available moisture. After mating, they lay 80 eggs at a time into a nest in the top soil. The eggs are white, spherical, about 3 mm (¼ inch) in diameter, and have a soft shell.

Management Guidelines

Biological Control. Several methods are available to control the brown garden snail. Biological control has been successful in recent years following the introduction of the decollate snail, *Rumina decollata*, a natural enemy native to North Africa and southern Europe. This predatory snail attacks and consumes young to half-grown brown garden snails. Since the early 1970s, the decollate snail has been colonized in many southern California citrus orchards, providing good control of the brown garden snail in many instances. To date, the California State Department of Fish and Game has not authorized the distribution of the decollate snail north of the Tehachapi Mountains.

The decollate snail is quite different in appearance from the brown garden snail. Its shell is elongated and about 2.5 cm (1 inch) long when mature. While the shell is growing, the top becomes brittle and breaks off. The decollate snail is self-fertile and lays about 500 eggs during its lifetime. The eggs are smaller than those of the brown garden snail and have a brittle shell. The decollate snail lives in the top soil and only emerges to feed or to escape heavy rail or irrigation water. Besides attacking the brown garden snail, the decollate snail feeds on decomposing leaves, fallen or bruised fruit, and emerging seedlings; it has not been observed to feed on sound fruit or citrus leaves.

The fastest way to establish the decollate snail in your orchard is to distribute 50 decollate snails per tree throughout the orchard, if such numbers are available. Several insectaries in southern Calfiornia are now rearing the decollate snail. Reduce the brown garden snail population at least 4 to 6 weeks before the release with a bait program or by mechanical removal. If you do not have enough snails to inoculate all trees, start in a cluster of core trees and provide an unbaited buffer zone between the expanding colony and the baited areas. The buffer zone should be at least two tree rows because the decollate snail will feed on snail bait. The rate of dispersal depends on the amount of moisture present; low volume and sprinkler irrigation are most conducive to snail movement and development. Light supplemental irrigations may be desirable during the establishment of a colony. After the colony has grown, you can transfer parts of it to other, nonbaited areas of the orchard to help the spread of the snail. Depending on the initial snail populations and on the environmental conditions, it can take from 2 to 5 years before the decollate snail can completely control the brown garden snail. Several insectaries and other sources in southern California have started to produce the decollate snail for release in citrus orchards. Check with your farm advisor for listings of these companies.

Monitoring. A wooden board (30 x 30 cm or 1 square foot), with cleats about 4 cm (1½ inches) high, is useful for monitoring the brown garden snail population or for removing large numbers of the snails. Place one to two boards per tree, depending on the size of the snail population, underneath the tree skirts in the core colonization area and check them about once a month. The brown garden snail will attach itself to the underside of the board, whereas the decollate snail rarely does so. To monitor the progress of your biological control program, observe the sizes of the snails; if full-grown brown garden snails are predominant on the boards, it means that the decollate snail is destroying most of the young snails. The presence of many young decollate snails on the soil indicates that the species is reproducing well. Once the number of decollate snails averages one-half the number of brown garden snails, the decollate snail is well established and will probably control the brown garden snail in 2 to 3 years. After checking the population, scrape the brown garden snails off the boards into soil pits and destroy them.

Baiting. A bait mixture of metaldehyde (2%) and methiocarb (1%) is the most effective pesticide for control of the brown garden snail. The time of baiting is critical; it should be done during a cool, damp period—when snails are very active—soon before dry, warm weather begins. Distribute the bait in open areas around the tree skirts; after feeding on the bait, the snails are too weak to move back into the shade and are killed by the sun. If cool, moist weather persists, many of the affected snails may recover. Carefully follow all instructions on the label. For other pesticide recommendations, see the most recent University of California *Treatment Guide for California Citrus Crops*. If you plan to release decollate snails, the waiting period after a baiting program depends on the soil moisture. Under sprinkler or low-volume irrigation, the toxins will break down faster than in drier soil, and a release program can start in 4 to 6 weeks.

Brown garden snails can be prevented from climbing into the trees by pruning the tree skirts up and applying a band of Bordeaux slurry around the trunks with a brush or garden sprayer once a year.

One of the most common diseases of citrus in all regions is Phytophthora gummosis. Sap oozing from trunk lesions is a typical symptom of this disease (page 114). Gumming is more pronounced in early spring, but it may be washed off during rain. Note that *Phytophthora* lesions extend high up on the trunk on susceptible cultivars.

Diseases

A number of microorganisms and environmental stresses can cause disease in citrus. Microorganisms, including fungi, bacteria, and viruses, produce biotic or infectious diseases. Abiotic diseases, such as genetic disorders, nutrient deficiencies, or adverse soil and weather conditions, may create conditions favoring the development of biotic diseases or produce disease symptoms of their own.

Several viruses, a few fungi, and one bacterium cause the major biotic diseases in citrus. Viruses and a mycoplasma, a viruslike organism, can infect all parts of a tree. Fortunately, careful propagation techniques can prevent many infections. Among the fungi, those infecting the roots and trunk are of greatest concern; fungal diseases of fruit are less important in the semiarid climate of California. Climate also limits the distribution of the only bacterial disease in California citrus—citrus blast.

The severity of a disease is influenced by the virulence of the pathogen, the genetic susceptibility and the growth stage of the host, and by environmental conditions such as temperature and humidity. The effect of a pathogen on a tree also depends on the site of infection. If the feeder roots, which are important in the uptake of water and nutrients, are diseased, the tree will suffer more readily than when some leaves and twigs are infected.

Certain regional differences in disease patterns can be observed. The viral disease tristeza is mainly restricted to the old citrus-growing areas of southern California. Great efforts in restricting the movement of propagating material and detecting and eradicating diseased trees have so far kept the San Joaquin and Coachella valleys free of tristeza. Stubborn disease, however, is a problem in the hot climate of these interior valleys. Citrus blast, a bacterial disease, affects citrus in northern California almost exclusively, where there are long, wet periods in winter and spring. Other diseases occur in most growing areas, and their damage potential depends more on the planted rootstock and scion, soil conditions, and water management practices than on climate or other regional differences.

Many diseases have ultimately the same effect, resulting in a declining tree with light green foliage, poor growth flushes, leaf drop, and twig dieback. Primary symptoms may appear on roots and trunks, on fruit, on leaves and twigs, or may affect the overall growth habit, size, and crop yield. Because primary symptoms help in identifying a disorder, the following section is organized according to where the major symptoms of a disease occur on the tree. Keep in mind that other pests, such as nematodes and gophers, or poor water management practices, may also be the cause of decline.

Monitoring and Diagnosis of Citrus Diseases

While walking through the orchard, look for weak trees and irregular growth patterns. Check problem areas, such as low-lying sites and fine-textured or shallow soil. Monitor more frequently when adverse weather persists or where trees are susceptible to diseases prevalent in the growing district. To diagnose a problem, look at the trunk for gumming, scaling, or shelling of the bark. You may have to scrape off weak or discolored bark, dig up some feeder roots, or remove soil from the crown or the main lateral roots extending outward to the edge of the canopy. A shovel, hatchet, chisel, knife, and hand lens are useful tools for diagnosing a problem. Some viral diseases, particularly tristeza, are difficult to identify by field symptoms; positive identification requires laboratory analysis.

Time of year influences what symptoms you may see; fungal and bacterial diseases of fruit and leaves cause most damage during the moist winter months. Diseases affecting the roots and trunks may become more apparent when trees are likely to be stressed during the hot, dry summer months.

Certain rootstock/scion combinations as well as individual rootstocks or scions are prone to incompatibilities or genetic disorders. Some of these diseases can be diagnosed by field symptoms, such as shelling or bark necrosis; others can only be identified by microscopic examination of tissues. Most of the disorders become evident long after the trees are in production, often after 10 to 15 years. At times, a mutation may occur in part of a tree, affecting fruit shape or color or the growth habit of leaves and twigs. Such somatic mutations are called chimeras or "sports."

Compare your field observations with the descriptions and photographs in this section. A combination of symptoms usually will identify a disease. Only the most common disease problems are included in this manual. If you find symptoms not described here, see the reference books on diseases listed in the appendix or consult your farm advisor.

Keep accurate notes of your observations and note the infected trees or those suspected of being diseased on a map of your grove. Positive diagnosis, especially of slowly developing viral diseases, is often not possible at first inspection. Repeated monitoring and mapping will help you confirm your diagnosis, follow the development of the disease, and make proper management decisions.

Prevention and Management

Prevention is the most economical and often the only way to manage diseases in citrus. Preventive measures include preplant soil preparation, planting resistant rootstocks and scions, careful water management and sanitation, and applying long-lasting fungicidal sprays. In addition, adequate fertilization, pruning, and control of other pests help maintain vigorous trees that are more resistant to infection than stressed, weak trees.

Disease management in citrus starts before planting. Fewer disease problems will occur if the site has deep, level soil with good drainage. Low areas and finely textured and shallow soils have the potential for water logging, which is detrimental to root health. Where hardpans obstruct water flow, ripping or backhoeing before planting helps improve water penetration. Subsurface tile drainage has been used successfully to avoid water logging. Carefully design the irrigation system and schedule irrigations according to the physical and climatic conditions of the site and the needs of the trees; for details, see the section on Irrigation. Good water management is essential for maintaining healthy roots and vigorous trees throughout the life of the orchard.

Selecting resistant rootstock and scion cultivars and fumigating a replant site are important means of preventing or reducing the effects of diseases. Resistant or tolerant cultivars are available for many common diseases, although no rootstock/scion combination is resistant to all important diseases and environmental stresses. The choice, therefore, depends on the prevailing disease problem and on soil conditions and climate. Farm advisors and nurserymen can provide information that helps you select the appropriate cultivars for your orchard conditions.

Using virus-free budwood and practicing good sanitation can prevent the spread of most virus diseases. Exceptions are tristeza and stubborn, which are spread by certain insect species. Clean propagation material is still valuable for reducing the incidence of these maladies. Virus-free budwood is available to nurserymen and growers in California through the Citrus Clonal Protection Program (CCPP), formerly called the Citrus Variety Improvement Program. The function of CCPP is to develop and maintain healthy, virus-free budwood of desirable, true-to-type citrus scion and rootstock cultivars. All distributed budwood comes from trees registered by the Nursery and Seed Service of the California Department of Food and Agriculture.

SYMPTOMS ON ROOTS

Three diseases commonly affect citrus roots in California: Phytophthora root rot, dry root rot, and Armillaria root rot. Because of their impact on the root system, the first visible symptom of all these diseases is usually a yellowing of leaves and a slow decline of the scion top. Phytophthora root rot destroys only the feeder roots; the other two diseases move into the woody parts of the root and crown.

Phytophthora Root Rot
Phytophthora spp.

Phytophthora root rot is a destructive fungal disease common in all citrus-growing areas. Rootstocks differ in their susceptibility to Phytophthora root rot; trifoliate orange, Alemow, and sour orange are the most tolerant stocks. Phytophthora fungi may also infect the trunk, causing gummosis (see page 114), or they may infect fruit, resulting in brown rot (see page 117).

Symptoms and Damage

As with other root diseases, Phytophthora root rot causes a slow decline of the scion top. The foliage turns light green or yellow and may drop, depending on the stage of infection. The disease destroys the feeder roots of susceptible rootstocks; it does not spread to woody roots or the crown. The pathogen infects the root cortex or rind, which turns soft and separates from the stele. If the destruction of feeder roots occurs faster than their regeneration, the uptake of water and nutrients will be severely limited and the tree will grow poorly and production will decline.

Seasonal Development

Phytophthora species are present in most citrus groves. They can survive adverse conditions as persistent spores in the soil. During moist conditions, these spores produce large numbers of motile zoospores, which can swim in water for short distances. Zoospores are the infective agents that are carried in irrigation or rainwater to the roots. The *Phytophthora* species causing root rot develop rapidly under the warm conditions of late spring and early summer.

Management Guidelines

Management of Phytophthora root rot focuses on preventing conditions favorable for infection and disease development. Good water management is essential in maintaining good root health and slowing the development of Phytophthora fungi. Provide adequate soil drainage and avoid over irrigating. Medium-textured, fairly deep soils are least subject to flooding or ponding and, therefore, to *Phytophthora* infections. (For details on soil and water management, see page 17.)

When replanting or establishing new plantings, choose resistant rootstocks where possible (Table 2, page 17), but also consider tolerance to other diseases, nematodes, and cold.

When planting or replanting in soil infested with *Phytophthora* or when a susceptible rootstock has to be used, fumigation may be feasible if no other adverse conditions persist. Carefully follow label instructions for dosages and waiting periods before planting trees.

If a tree growing on a susceptible rootstock looks stressed, dig up a few handfuls of soil and check the health of the feeder roots. Healthy roots are firm, whereas roots affected by root or foot rot have a soft rind that easily slides off the stele when pinched. Often, however, disease symptoms are difficult to distinguish from nematode damage; only a lab analysis can provide positive identification (see Nematodes section). If the destruction of feeder roots is minimal, corrective actions may halt the decline. Once many feeder roots are diseased, the tree will continue to decline and become unproductive. Corrective actions may include increasing irrigation intervals, switching to alternate middle irrigation or a different irrigation system, and installing subsoil tiles.

Dry Root Rot

Dry root rot causes injury similar to the injury from Phytophthora root rot. Although the disease is normally a chronic problem affecting only a few scattered trees in a grove, it can develop into an epidemic in some orchards. The exact cause of dry root rot has not been established, but a *Fusarium* fungus is most often isolated from diseased wood. All common rootstocks, including trifoliate and Troyer citrange, which are fairly resistant to *Phytophthora*, are susceptible to dry root rot.

Symptoms and Damage

The general symptoms are similar to those caused by other agents that damage the roots or girdle the trunk. These include reduced vigor, dull green leaf color, poor new growth, and twig dieback. If extensive root damage occurs, the leaves suddenly wilt and dry up on the tree.

Damage usually starts in larger roots and spreads into the crown. Patches or large areas of bark at the crown show a moist, dark decay, which later dries and adheres to the wood. The wood below the dead bark is hard, dry, and stained grayish brown to purple. Unlike Phytophthora gummosis, dry root rot does not produce gumming, and the lesion extends deep into the wood. Once the crown region is girdled, the tree collapses.

Seasonal Development

Dry root rot organisms often infect a tree through the crown or larger roots that have been injured by *Phytophthora* fungi, tillage, gophers, or overdoses of fertilizer, herbicides, or nematocides. The initial infection may occur at the time of planting or at any point during the trees's lifetime, but aboveground symptoms may only appear several years after the initial infection when the crown region has been girdled.

The pathogens most commonly associated with dry root rot are *Fusarium solani* and various bacteria; sometimes parasitic and saprophytic fungi are also present.

Management Guidelines

Good orchard management, especially careful irrigation, is essential for preventing dry root rot. If the soil around the tree crowns and roots is saturated for long periods of time, the chances for injury and subsequent fungal infections increase. When establishing furrows, provide berms along the trees so that the crowns are protected from the water. When using sprinklers, direct the water away from the trunks.

Injury to roots and trunk, especially during the cool and wet season, provide entry wounds for infection. Carefully follow label instructions for applying fertilizer, herbicides, and other pesticides to prevent injury. Before fertilizing young trees, wait at least 6 weeks after planting or until the trees show new growth. When carrying out cultural operations, such as controlling weeds or removing sucker shoots, avoid injury to the lower trunk, crown, and feeder roots in the top soil.

Check regularly for signs of Phytophthora root rot or vertebrate damage that may provide entry sites for dry root rot. Take actions to prevent Phytophthora gummosis or to halt the disease. Control gophers when you detect their activity in the orchard.

If you suspect a dry root rot infection, dig all the way around the tree because the decay may be underneath the crown roots or on one or more of the main lateral roots. You can often halt disease spread by exposing the crown region, enabling it to dry. Prune up the tree skirts and remove the soil from the crown region. Correct any adverse soil conditions, such as poor drainage, that may contribute to dry root rot infections. Remove trees that have become unproductive because of severe infection.

Armillaria Root Rot
Armillariella mellea

Armillaria root rot, also known as oak root fungus disease, can occasionally damage and kill citrus trees. The Armillaria root rot fungus is native to California oak trees and affects many tree crops planted in former riverbeds, floodplains, or other land subject to overflow.

Symptoms and Damage

Symptoms may not develop until after the disease is well established. The first symptoms of Armillaria root rot

Dry root rot affects the roots and crown region; it usually does not extend into the trunk above the bud union. The bark has been removed to show the stained wood below and the healthy, greenish white wood above.

The *Armillariella* fungus forms whitish mycelial plaques between the bark and wood in the crown region.

are poor growth or dieback of shoots, small yellowing leaves, and premature leaf drop. The pathogen invades the roots and crown, eventually girdling the crown region and destroying the entire root system.

The parasitic fungus spreads by root contact or through rhizomorphs, black strings of fungal mycelia, which can grow short distances through the soil and contact and penetrate citrus roots. From the infection site, the fungus invades lateral roots and the crown region, where it spreads as white mycelial plaques between the bark and wood. This distinguishes *Armillariella* from wood-rotting fungi, which grow on the outer surface of the bark.

Seasonal Development

The Armillaria root rot fungus can survive for many years in dead or living roots of fruit and nut trees, oak, aspen, and many other species. The disease often occurs along former stream beds and near creeks where the soil is moist and where dead roots or stumps harboring the fungus may be buried in the soil. The fungus requires cool, moist soil conditions for development and spread; it is therefore rarely a problem in desert areas.

In winter, *Armillariella* often forms clusters of mushrooms at the base of infected trees a few days after a rain. The mushrooms vary greatly in size, shape and color, but they all have a veil or ring on the stalk just under the cap and shed white spores.

Management Guidelines

Management of Armillaria root rot relies primarily on preventing infection of new trees. Once infection is apparent, it is very difficult to save a tree. You can reduce the chances of infection by carefully preparing planting sites for new orchards and by practicing good sanitation and early detection.

Avoid a planting site likely to be infested with *Armillariella*. To prepare infested sites for replanting, remove and destroy large roots, and fumigate the site. See your farm advisor for advice on fumigation materials, application, and precautions. If there are infected trees in your orchard, remove and burn them and the neighboring, apparently healthy trees; once symptoms appear on a tree, the disease has probably already spread to the roots of the surrounding trees.

SYMPTOMS ON THE TRUNK

Various diseases and disorders produce symptoms on the trunk of citrus trees. Common symptoms are shelling and scaling of the bark and development of lesions and gumming. Serious damage to the trunk disrupts transport of water and nutrients and results in overall decline of the tree. Phytophthora gummosis is the most common disease of the trunk. Other diseases are caused by the exocortis viroid, psorosis virus, genetic disorders (shell bark and dry bark), bud union disorders and sunburn.

Phytophthora Gummosis
Phytophthora spp.

When *Phytophthora* fungi infect the trunk of citrus trees, the disease is called Phytophthora gummosis. All scion cultivars are susceptible to infection under certain environmental conditions. If the rootstock is susceptible to Phytophthora infection, the disease may spread into the crown and woody roots; it is then sometimes referred to as foot rot.

Symptoms and Damage

An early symptom of Phytophthora gummosis is the sap oozing from small cracks in the infected bark, giving the tree a bleeding appearance. The gumming may be washed off during heavy rain. Gummosis affects the bark and the cambium but does not destroy the wood, although a thin layer of wood may be impregnated with gum. The bark stays firm, dries, and eventually cracks and sloughs off. Lesions spread around the circumference of the trunk, slowly girdling the tree. Decline may occur rapidly within a year, especially under conditions favorable for disease development, or may occur over several years.

Secondary infections often occur through lesions created by *Phytophthora*. These infections kill and discolor the wood deeper than gummosis itself; an example is dry root rot (see p. 112).

Seasonal Development

Phytophthora fungi are present in almost all citrus orchards. Under moist conditions, the fungi produce large numbers of zoospores, which are splashed by rain or irrigation water onto the tree trunks. The *Phytophthora* species causing gummosis develop rapidly under moist, cool conditions. Hot summer weather slows disease spread and helps drying and healing of the lesions.

Management Guidelines

Management of Phytophthora gummosis focuses on preventing conditions favorable for infection and disease development. When establishing a new orchard, carefully check the lower trunk and rootstock of new trees for any symptoms of gummosis. When trees are wrapped in burlap, open and inspect a representative sample (at least 10% of the trees). Try to reach an agreement with your nursery owner allowing you to reject trees that show any signs of gummosis. If you have susceptible rootstocks, provide good aeration of the crown region. Where necessary, remove the soil to where large roots branch off and provide drainage for irrigation or rainwater.

When planting or replanting in soil infested with Phytophthora, or when a susceptible rootstock has to be used, fumigation may provide some control. Carefully follow label instructions for dosages and waiting periods before planting the trees. Plant trees on a berm or high enough so that the first lateral roots are just covered with soil. This will help keep the crown region dry during irrigation or rain. The bud union should also be high enough above the soil line to avoid infection.

Late stages of Phytophthora gummosis are distinct, but early symptoms are often difficult to recognize. Yet early detection and prompt management actions are essential for saving a tree. If 50% or more of a trunk or crown region on a mature tree is girdled, it is more economical to replace the tree than to try to control the infection.

Inspect your orchard several times a year for disease symptoms. Look for signs of gumming on the lower trunk and crown, and for soil buildup around the crown. Wrappers on young trees should be lifted or removed for inspection. When you detect gum lesions, check soil and drainage conditions. Correcting any soil or water problems is essential for a recovery.

In addition to improving the growing conditions, you can halt disease spread by scraping off the diseased, dark bark and a buffer strip of healthy, light brown to greenish bark around the margins of the infection. Allow the exposed area to dry out. You can also scrape the diseased bark lightly to find the perimeter of the lesion and then use a propane torch to burn the lesion and a margin of 2.5 cm (1 inch) around it. Recheck frequently for a few months and repeat the procedure if necessary. A material is also available to treat nonbearing citrus; for fungicide recommendations, see the latest University of California *Treatment Guide for California Citrus Crops*.

Exocortis
Exocortis Viroid

Exocortis is of minor importance in California today because strict regulations on budwood sources have kept new plantings largely free of this viroid disease. Exocortis is widespread in older plantings, but it is a mild disease that causes only moderate stunting and some loss of productivity. The viroid is easily spread on infected budwood and contaminated propagation tools.

The characteristic symptom of exocortis is the shelling of susceptible rootstocks. The viroid kills the bark, which dries, cracks, and may lift in thin strips. Some droplets of gum often appear under the loose bark. The extent of shelling varies. Sometimes lesions develop slowly and shelling is restricted to a small area for several years; at other times, the disease develops rapidly over the entire rootstock. Infected trees rarely die, but growth is retarded and productivity slowly declines. Among the commonly used rootstocks, trifoliate is most affected by exocortis (Table 2).

For planting or replanting, it is best to use viroid-free budwood available to nurseries or growers through the Citrus Clonal Protection Program. These clean bud lines can be grafted onto susceptible rootstocks. A few desirable scion cultivars still carrying a mild exocortis viroid should be grafted only onto tolerant rootstocks. It is best to remove infected trees from the orchard because pruning clippers and saws can transmit exocortis unless thoroughly disinfected with hypochlorite; heat does not kill the viroid.

Psorosis
Psorosis Virus

Psorosis is a virus disease often found in old citrus plantings. Infected trees, mostly orange and grapefruit, slowly decline; main scaffold branches die so that the trees become unproductive.

The most distinguishing field symptom is scaling and flaking of the bark on the scion cultivar. During early stages, patches of bark on the trunk or scaffold branches show small pimples or bubbles, which later enlarge and break up into loose scales. Gumming often appears around the margins of a lesion. In an advanced stage, deep layers of bark and the wood become impregnated with gum and die.

The psorosis virus is transmitted in infected budwood or possibly with contaminated grafting tools. Occasionally, the disease spreads through root grafting from an infected to a healthy tree. Seeds of some citrange cultivars are known carriers of the virus.

As with other viral diseases, the use of virus-free budwood is the major method for preventing damage from psorosis. The Citrus Clonal Protection Program provides budwood free of major viral diseases to nurserymen and growers. Where an old tree shows symptoms, scraping away the infected bark area will stimulate the formation of wound callus and result in temporary recovery. Generally, a psorosis-infected tree will produce less, and replacement is the best management option.

Other Trunk Diseases and Disorders

Shell Bark and Dry Bark

Shell bark is apparently an inherited disorder that occurs on lemons. Nucellar seedlings are less severely affected than old lines of lemons. Symptoms begin to appear after 10 years on Eureka lemon and even later on certain

Lisbon lines. Small areas of the outer bark die, crack, and loosen in long strips. Shelling usually moves from the bud union upward, but sometimes starts in the crotch of scaffold branches. Lesions enlarge very slowly. On Lisbon, the disease usually does not cause noticeable decline; on Eureka, shelling may be so extensive that the required regeneration of bark tissue weakens the tree.

If the disorder extends into the inner bark and cambium, the disorder is called dry bark; the affected areas dry and crack slightly but do not shell because no new tissue is generated underneath. Secondary decay organisms may invade the lesions and aggravate the decline. If dry bark extends around most of the tree's circumference, food transport is severely limited and the tree will die.

In winter, *Armillariella* often forms clusters of mushrooms at the base of infected trees after a rain.

The characteristic symptom of exocortis is the bark shelling on susceptible rootstocks.

Psorosis causes a scaling and flaking of the bark on susceptible scion cultivars.

Certain rootstock-scion combinations develop delayed bud union disorders. Here a small square of the outer bark is cut out to show the crease that forms at the bud union because of the uneven growth between rootstock and scion. The crease compresses the food-conducting vessels, slowly girdling the tree.

As a preventive measure, use only budwood from tolerant parent trees. In old orchards, severe pruning of diseased trees helps stimulate new growth and regeneration of affected bark tissue, allowing at least temporary recovery.

Bud Union Disorders

Certain scion-rootstock combinations show an incompatibility reaction, which may appear shortly after grafting or may take 10 to 20 years to develop. An example of delayed incompatibility is bud union crease. A crease or fold forms at the bud union, and with increasing overgrowth, the food-conducting tissue is compressed, resulting in a slow girdling of the tree.

When planting or replanting, avoid susceptible combinations. For example, Frost nucellar navel on trifoliate, lemon on Cleopatra mandarin, Eureka lemon or Satsuma mandarin on Troyer, and certain scion lines on sour orange develop bud union overgrowths. For more information, contact your farm advisor's office.

Sunburn

The bark of young trees or severely pruned trees sunburn easily. Sunburned bark appears dried and often cracks and sheds. The injury can provide entry sites for decay organisms. To protect tree bark against sunburn, paint young tree trunks with whitewash or use tree wrappers. After a severe pruning, protect exposed limbs with a whitewash until the canopy grows back. Intensive sun radiation may also blemish fruit (see page 119).

SYMPTOMS ON FRUIT

Many pathogens and environmental conditions damage citrus fruit. The most important is brown rot caused by *Phytophthora* fungi. Other fungal diseases affecting fruit in California include Alternaria rot, Septoria spot, Anthracnose tearstain, blue and green mold, and Botrytis rot. Adverse weather including frost, cold and wet, wind, and sunburn may also injure fruit. Other disorders are caused by pesticide injury or new mutations that produce chimeras.

Brown Rot
Phytophthora spp.

Under wet, cool conditions, *Phytophthora* fungi can cause brown rot on fruit. Damage from the disease may occur each winter in most citrus-growing areas of California. Trees can show symptoms of brown rot alone or in combination with symptoms caused by Phytophthora

Under cool, wet conditions, brown rot often develops on fruit. The infection starts as firm, water-soaked spots that develop into large, tan to olive brown areas that emit a pungent odor. The fruit in this picture are in an early stage of infection.

Alternaria infects the navels of navel oranges, especially when the navel is split. Usually it also causes a premature color change.

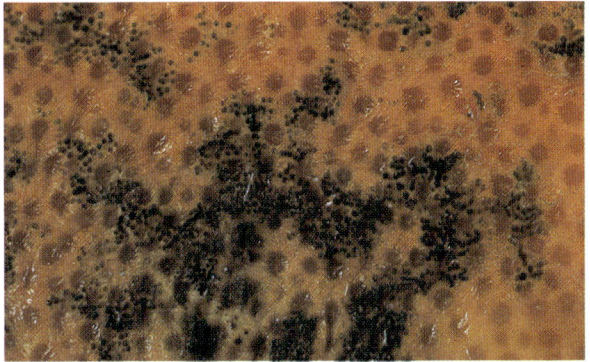
Fruit infested with Septoria develop small, light tan to reddish brown pits. Dark brown to black fruiting bodies growing in the pits distinguish a Septoria infection from copper injury.

gummosis (see page 114) or Phytophthora root rot (see page 112).

Brown rot develops mainly on fruit growing near the ground. *Phytophthora* spores from the soil are splashed onto the tree skirts during rain storms, and infections develop under continued wet conditions. Symptoms appear primarily on mature or nearly mature fruit. Initially, the firm, leathery lesions have a water-soaked appearance, but they soon turn soft and have a tan to olive brown color and a pungent odor. The infection progresses over the fruit surface, but not beyond the albedo. Infected fruit eventually drop. Fruit in the early stage of the disease may go unnoticed at harvest and infect other fruit during storage. Occasionally, twigs, leaves, and blossoms are infected, turning brown and dying.

Management Guidelines

The only management option for brown rot is prevention. Apply a preventive copper fungicide in October or November before or just after the first rain. Generally, one outside coverage spray provides protection throughout the wet season, but when rainfall is excessive, you may have to repeat the spray in January or February. Covering the tree skirts about 4 feet up is usually sufficient and does not harm natural enemies. Spraying the ground underneath the trees also reduces brown rot infections.

The preferred chemical control for brown rot is Bordeaux, a mixture of copper sulfate, hydrated lime, and water. The spray adheres well to foliage, even through repeated rains. It releases copper ions, the active ingredient, slowly during moist conditions when zoospores are most active. The ratio of copper in the mixture depends on the intended use and the danger of copper damage. Where copper toxicity is a problem, such as in certain areas of Riverside, Ventura, and Kern counties, a greater proportion of hydrated lime will slow the release of free copper, or captan can be substituted. You can also replace a certain percentage of copper with zinc, which is somewhat toxic to *Phytophthora* and provides some of the nutritional requirements for zinc. A copper/zinc/lime mixture also protects against Septoria spot.

Neutral copper fungicides, also called low soluble or fixed copper fungicides, can also be used to control brown rot.

For recommendations in preparing and applying Bordeaux and other fungicides, see the latest University of California *Treatment Guide for California Citrus Crops*.

Alternaria Rot
Alternaria citri

Alternaria rot is a fungal disease that affects mainly navel oranges and lemons. On navel oranges, the disease is also called black rot, and results in dark brown to black, firm spots or areas at the stylar end or in the navel. If you cut the fruit in half, you can see the rot extending into the core. Fruit infected with *Alternaria* change color prematurely. The decay is softer on lemons than on oranges and develops mostly during storage. Healthy, good quality fruit are more resistant to Alternaria rot than stressed or damaged fruits, especially oranges with split navels. Preventing stress can reduce the incidence of splitting and Alternaria rot.

Septoria Spot
Septoria spp.

The *Septoria* fungus causes spotting of Valencia oranges and occasionally of lemons and grapefruit. It occurs in the San Joaquin Valley and interior districts of southern California during cool, moist weather.

Infections begin when Septoria spores are spread throughout the tree in dew or rain water. On fruit, the symptoms appear as small, light tan to reddish brown pits, 1 to 2 mm in diameter. Dark brown to black fruiting bodies often grow in the small depressions, which usually do not extend beyond the oil-bearing tissue. The spots are much more conspicuous after the fruit has changed from green to yellow or orange. A tearstaining often accompanies the spotting. The rind damage lowers the grade and results in culling. On lemon, the small spots develop into large, brown blotches during storage. Anthracnose or Alternaria fungi often occur as secondary infections on Septoria lesions. Septoria spot may be confused with copper injury, except that the pits caused by copper do not have fungal fruiting bodies.

Septoria may cause similar spotting on leaves or twigs that are weakened or damaged by frost or pests. Mature trees with many dead twigs are more likely to become infected with *Septoria* than young, vigorous trees.

To control Septoria spot, apply a preventive copper spray in late fall or early winter, just after the first rain. A neutral copper fungicide, which does not contain lime, is preferred in areas with parasite populations. A copper spray also protects against brown rot. Thorough coverage of the entire tree is essential, and in years with heavy rainfall, retreatment may be necessary. For pesticide recommendations, see the latest University of California *Treatment Guide for California Citrus Crops*.

Other Diseases and Disorders of Fruit

Anthracnose Tearstain
Collectotrichum gloeosporioides

Anthracnose tearstain may blemish the rind tissue of mature Valencia and navel oranges, grapefruit, and occasionally lemon. The disorder affects mainly fruit on stressed trees with old, dead wood. The Anthracnose fungus usually first infects weakened twigs. During wet or foggy weather, Anthracnose spores drip onto fruit, where they infect the rind and leave dull, reddish green streaks. The stain cannot be washed off, but the disorder is generally not severe enough to require preventive actions.

Anthracnose tearstain often occurs with Septoria spot. The Septoria fungus itself and possibly certain environmental conditions may also cause tearstaining.

Blue and Green Mold Disease   *Penicillium* spp.

Blue and green mold disease is mainly a problem during storage but may occur on injured fruit in the field. Although early infections are almost impossible to detect, the disease is easily recognized once the whitish mycelium and the blue or green spores appear. Blue and green mold usually occur together, although the blue mold often overgrows the green mold.

To reduce infection with blue and green molds, do not pick wet fruit and handle fruit carefully during picking and transport. No field treatments are recommended for control.

Botrytis Rot   *Botrytis cinerea*

Botrytis rot is sometimes a problem on coastal lemon and occasionally on Valencia oranges. During prolonged wet periods, the *Botrytis* fungus, also called gray mold, infects blossoms and also twigs (see page 119). Petals carrying the fungus often spread the disease when they contact small fruit. At the point of contact, the fungus stimulates cell growth, so that ridges develop. Mature Valencias develop a soft decay at infection sites.

No treatment guidelines are available for Botrytis rot. General preventive measures, such as avoiding mechanical or chemical injury, protecting against frost and brown rot, and pruning regularly, help reduce the incidence of Botrytis rot.

Chimeras

If a mutation occurs in a branch or twig and survives, it may develop into a new twig with characteristics different from the rest of the tree, called a chimera or "sport." New characteristics may affect the color of the rind or pulp or the shape of the fruit, often resulting in bizarre forms, for example, the so-called cock's comb. The mutation may affect the shape, size, or color variegation of leaves (see page 119). At times, a chimera has produced an improved crop; some of today's cultivars were propagated from chimeras. Generally, however, the sports are of inferior quality and should be avoided as propagation material.

Frost Damage

A subtropical plant, citrus is particularly susceptible to frost damage, but the susceptibility varies with the physiological state of the tree, the rootstock cultivar (Table 2), and the scion. For example, Eureka lemon and grapefruit are among the most tender cultivars, whereas sweet orange, mandarin, and Meyer lemon are more cold hardy.

Fruit is most susceptible to frost damage, although twigs and leaves may be affected (see page 119). On fruit, the oil from ruptured oil cells corrodes the rind surface, causing watery, brownish specks or pits called ice marks. Frost damage mainly appears on the outside of fruit exposed to radiation frost, and the pulp underneath the ice marks ultimately dries. Intense damage may occur without the rind markings. Ice marks also provide entry sites for decay organisms. Certain cultural practices can help reduce the impact of frost (see page 23).

Rind Stipple of Grapefruit

Rind stipple of grapefruit is associated with a particular weather pattern. During prolonged periods of cold, wet weather, the rind of grapefruit develops small, brownish pits. The pits often have concentric rings around them or they coalesce into larger, irregular lesions. If the fruit is injured while green, a green halo appears around the pits and persists until the fruit has turned yellow. The symptoms occur mostly on exposed fruit on the north side of the tree. The rind damage, which seems to affect only grapefruit, lowers the grade of the fruit. No treatment is available.

Sunburn

Intensive sun radiation may blemish mature fruit. The side of fruit continuously exposed to the sun develops brownish, leathery areas; some fruit become slightly lopsided. Application of whiting agents can reduce sunburn but is usually not economical.

Wind Injury

In areas with persistent winds, fruit on the outside canopy may be scarred where twigs and thorns rub against the rind. During strong winds, fruit also drop from the tree and leaves can be damaged (see page 119). Windy locations should be protected by windbreaks.

Spray Injury

If pesticide sprays, in particular herbicides and fungicides, are applied incorrectly, damage to fruit may occur. If paraquat drifts or is sprayed onto green fruit, brown necrotic spots develop at the point of spray contact; these symptoms do not appear on ripe fruit. Weed oil and dinoseb produce necrotic spots on green or ripe fruit. Fungicidal sprays containing copper may cause pitting on fruit or leaves which resemble a Septoria infection but do not have fungal fruiting bodies. Certain areas in southern California and Kern County are more likely to suffer copper injury than other citrus-growing regions. In high risk areas, use less copper in your spray or soften neutral copper fungicides with lime.

120 DISEASES

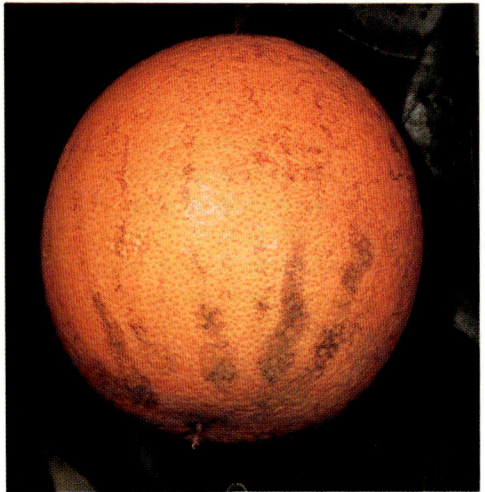

An Anthracnose infection typically results in a tearstained pattern of dull, reddish green streaks.

Blue green mold disease (left) starts as small, watery fruit lesions, which enlarge and produce white mycelia and blue or green spores. In coastal areas, Botrytis rot, another fungal disease, may infect blossoms or young fruit in the spring (right). Its mycelia spread under cool, moist conditions, killing the blossoms or fruit.

A mutation may occur in a branch, resulting in a chimera or sport, which produces abnormal fruit. Here segments of the rind are enlarged.

A severe frost damages epidermal cells, and the pulp begins to lose water. Internal damage begins to appear after one to several weeks, depending on weather conditions. Fruit at left is frost damaged.

Grapefruit are susceptible to rind stippling during prolonged periods of cold, wet weather.

During persistent winds, fruit may rub against twigs or thorns and develop scars, which are sometimes confused with citrus thrips damage.

SYMPTOMS ON LEAVES AND TWIGS

Several pathogens and various environmental conditions, toxic chemicals, and genetic disorders may cause leaves and twigs to wilt, dieback, or become discolored without affecting other parts of the tree. Most of these diseases do not cause serious long-term losses in citrus; but changes in management practices can often reduce their occurrence and improve the general health of trees. Mineral deficiencies and toxicities can also cause leaf and twig symptoms. Correct these problems as soon as they appear. Remember that diseases and other pests affecting roots, trunks, and overall tree growth often cause symptoms on twigs and leaves as well; if other parts of the tree also appear to be affected by the disease, consult the appropriate sections of this book.

Citrus Blast *Pseudomonas syringae*

Citrus blast is restricted mainly to citrus-growing areas in northern California, where wet, cool, and windy conditions during winter and spring favor development and spread of the blast bacterium. Leaves and twigs of oranges and grapefruit and the fruit of lemon are most susceptible to infection.

The bacterium infects small injuries caused by thorn punctures, wind abrasions, or insect feeding. Infections usually start as black lesions in the leaf petiole and progress into the leaf axil. Once the petiole is girdled, leaves wither, curl, and eventually drop. Entire twigs may die back. The damage is most severe on the south side of the tree, which is exposed to the prevailing winter winds. Diseased areas are covered with a reddish brown scab. Infections result in small, black spots on the fruit; thus the disease is also called black pit.

Preventive treatment against citrus blast alone is generally not economical, but sprays against brown rot or Septoria may provide some protection against citrus blast. Certain cultural practices can reduce the incidence of citrus blast: Planting windbreaks and using bushy cultivars with relatively few thorns help prevent wind injury; pruning out dead or diseased twigs in spring after the rainy period reduces the spread of the disease; and scheduling fertilization and pruning during the spring or early summer prevents excessive new fall growth, which is particularly susceptible to blast infection.

If weed oil or dinoseb is accidentally sprayed on fruit, necrotic spots develop at the site of contact.

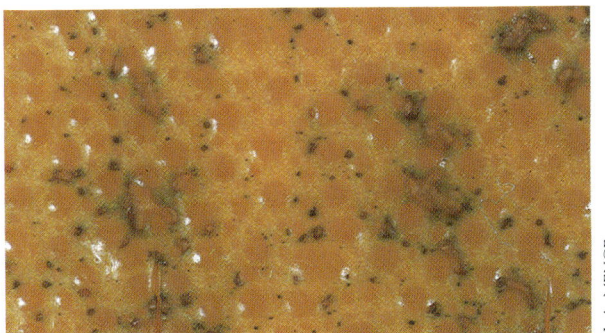

Copper sprays may cause fruit pitting under certain environmental conditions. The pits resemble Septoria infections but are without the fungal spore cases.

A citrus blast infection starts in leaf petioles and progresses into leaf axils. Infected leaves curl, dry, and eventually drop.

Botrytis Rot *Botrytis cinerea*

Botrytis rot is a problem mainly on lemon fruit (see *Symptoms on Fruit*), but the fungus may also infect twigs and small branches of lemon trees. The fungus infects through injuries, forming gray, velvety mats of fruiting bodies on the surface. Infected twigs may die back several inches. General preventive measures, such as avoiding mechanical or chemical injury, protecting against frost and brown rot, and pruning regularly, help reduce the incidence of Botrytis rot.

Chimeras

At times, a mutation in part of a tree results in the growth of one or several twigs with characteristics different from the rest of the tree. The leaves on these twigs may have a different shape, size, or color variegation. Or the mutation results in the development of multiple buds, creating a bunch growth or "witch's broom." Certain mutations also affect the shape or color of fruit (see page 122). Chimeras or "sports" are usually of inferior quality and should not be used for propagation. Prune sports that obstruct normal growth or interfere with harvest.

Twig Dieback

Twig dieback is periodically a problem on citrus trees in most growing districts. Many factors may contribute to the damage, for example, a poor root system, a mild scion-rootstock incompatibility, or weather stress. An abrupt twig dieback sometimes occurs in the San Joaquin Valley in the spring, when the largely inactive roots in the cool soil cannot meet the suddenly increased water demand of the canopy, resulting in water stress despite ample soil moisture. The stress causes gum formation in the conducting tissue of the twigs, aggravating the water transport problem. Leaves and twigs may die immediately, and leaves dry up on the tree, or they may die some weeks later, often at the time of another hot period. On navel orange trees, this type of dieback often occurs after the fruit has been picked; while on the tree, the fruit acts as a water reservoir for the leaves, alleviating water stress.

Another type of dieback affects limbs and branches and usually follows a dry period the preceding fall. No gum forms in the conducting tissue, and leaves do not drop until twigs or leaves are completely dry. Poor root health aggravates this condition.

Wind Injury

Strong, persistent wind can damage citrus by dehydrating the leaves or even stunting the growth of young trees. Drying winds such as the Santa Ana cause bronzing, pitting, and curling of leaves, mostly on the outside of the canopy exposed to the wind. In certain coastal districts, chilling winds blow from the ocean and inhibit the normal growth of citrus. Strong winds may also scar fruit (see page 122).

Windy locations require wind breaks for protection. A natural windbreak can be a row of fast growing, tall trees, such as eucalyptus; individual shelters of burlap or wood frames may be suitable for young trees.

Mesophyll Collapse

Mesophyll collapse on leaves can be a problem under certain environmental conditions. Soft interior leaf tissue between the veins (mesophyll) collapses and becomes translucent. Sometimes leaves dry completely and turn brown. These symptoms appear when the trees are unable to supply enough water to the leaves either because of low soil moisture or hot, dry winds. A poor root system, saline soil, and heavy mite feeding will compound the stress on the trees.

Frost Damage

Citrus is susceptible to frost damage, but the susceptibility varies with the physiological state of the tree and the cultivar (Table 2). Frost-damaged leaves or twigs become water soaked, wither, and turn dark brown to black. A frost may kill young or regrafted trees but rarely mature trees. Certain cultural practices, such as avoiding pruning or fertilizing during late summer and providing protection during critical injury times, can reduce the impact of frost (see Frost Protection, pp. 23).

Spray Injury

Pesticide sprays, such as oils, growth regulators, fungicides, and herbicides, may injure citrus trees if used incorrectly. A persistent oil film interferes with physiological functions of the tree, reducing sugar production and color development of the fruit or, in severe cases, resulting in fruit and leaf drop. Growth regulators, such as 2,4-D, cause leaf curling and chlorosis if applied at the wrong concentration or the wrong time.

Fungicides containing copper may cause a pitting of leaves and fruit. On leaves, pitting begins on the lower leaf surface and may extend through to the upper surface. Certain areas in southern California and Kern County are more likely to suffer copper injury than other citrus-growing regions. In high risk areas, use less copper in your spray or add lime to neutral copper fungicides.

Postemergence and preemergence herbicides can injure citrus by direct contact with foliage or fruit or by uptake through the roots. When roots take up too much of a preemergence herbicide, symptoms appear above ground mainly on the leaves. Bromacil damage causes extensive vein clearing, leaving a net pattern of bleached veins. Simazine results in a clearing of interveinal areas, not unlike symptoms caused by nutritional deficiencies. When too much diuron is taken up, major leaf veins turn yellow, and the interveinal areas remain green. Depending on the amount absorbed, any of the three herbicides may turn entire leaves brown. Localized damage on leaves or fruit may occur when a contact herbicide is sprayed onto the tree skirts. If paraquat is accidentally sprayed on or drifts onto citrus trees, leaves may drop. Glyphosate may deform new growth, darken foliage, and abort flowers. (For symptoms on fruit caused by contact herbicides, see page 122.)

Mineral Deficiencies and Toxicities

Mineral deficiencies occur because of low amounts in the soil or conditions that reduce the plant's ability to absorb them, such as excess moisture, low temperature, and root damage from biotic diseases. An excess of minerals in the soil or irrigation water may be toxic to the plant. The lack or excess of certain minerals usually appears first in leaf symptoms but eventually affects fruit size, quality, or yield.

In California citrus, only a few nutrient deficiencies are of concern. Nitrogen and zinc are most commonly deficient, manganese and magnesium are less frequently lacking, and iron is only occasionally deficient. Zinc and manganese deficiencies are common on alkaline soils, and zinc deficiency is most severe on sandy and sandy loam soils. Iron deficiency often results where soils drain poorly or have a high lime content. Toxicities from minerals are less common than deficiencies, although occasionally boron causes toxic symptoms where soils or irrigation water contains high levels of this mineral.

Many common deficiencies or toxicities produce distinct symptoms. At times, however, a particular disorder is difficult to identify because the symptoms are masked by other nutritional disorders or biotic diseases. A leaf analysis can verify field identification and provide information on the general nutritional status (see page 22).

Symptoms

You can often recognize a deficiency by the location of its symptoms on the tree and the pattern of leaf discoloration. Zinc and manganese deficiencies are most pronounced on young shoots, with the effects of zinc deficiency most noticeable on the south side and those of manganese deficiency on the north side of trees. When the deficiency is mild, the symptoms disappear later in the season. Magnesium deficiency is most prevalent on mature leaves in late summer or fall. Iron deficiency is manifested typically on the lower half or skirt of the tree, affecting leaves on the outside as well as the inside of the canopy.

Leaves deficient in zinc remain small and have pale to white, chlorotic blotches between the veins; often dark green, round spots are scattered throughout the leaf. Manganese deficiency causes a fine mottling of leaves, usually young leaves, with pale to yellowish interveinal areas and dark green, sharply contured vein areas. The deficiency often disappears as leaves mature. At times, manganese and zinc deficiencies occur together and symptoms are not as distinct. Leaves deficient in magnesium develop chlorotic blotches along the midrib. The blotches enlarge progressively, leaving only a center V-shaped, dark green area. In severe cases, they turn yellow and drop. When iron is deficient, leaves slowly lose their green color, eventually becoming yellowish to cream colored. Leaf veins remain green except in very severe cases.

Nitrogen deficiency does not produce distinct symptoms; a pale green to yellowish leaf color may also be due to root diseases or other pest problems. All deficiency symptoms except those of magnesium, which usually show up in late summer or fall, are most apparent during the winter and early spring.

Excess boron causes leaf tips and margins to turn yellow. The chlorosis progresses inward, and brown, dried gum spots appear on the underside of leaves. Eventually the tips and margins turn brown and wither. Symptoms are most apparent on older leaves in late summer and fall. Lemon and grapefruit are more susceptible to boron toxicity than oranges.

Treatment

Foliar sprays can correct zinc and manganese deficiencies. For recommendations, see the latest Agricultural Sciences Publications Leaflet 2903, *Treatment Guide for California Citrus Crops*. Iron deficiency is more difficult to correct. Where water logging is the main cause of the deficiency, a change in irrigation practices can alleviate the problem. Chelate products applied to the soil near the tree have also been used with some success. See your farm advisor for details.

Where you suspect boron toxicity, have your irrigation water checked. If the level of boron is high (more than 0.5 ppm), the best remedy is to switch to another source of water. Where this is not possible and the injury is severe, a more tolerant crop should be grown. Where injury is moderate, frequent, heavy irrigation can wash the excess boron below the root zone. Additional nitrogen fertilizer, especially calcium nitrate, and some liming of acid soils may also alleviate boron toxicity.

At times, a mutation in a branch may result in a chimera, that is, in leaves of abnormal shape or size or, as shown here, a color variegation.

When trees are unable to provide enough water to the leaves, a condition called mesophyll collapse may develop. The soft tissue between leaf veins collapses and becomes translucent or light brown.

Uptake of the herbicide bromacil may cause extensive vein clearing, resulting in a net pattern of bleached veins.

Frost may kill tender leaves and twigs.

Damage from simazine is apparent as interveinal chlorotic blotches or areas, depending on the amount of absorbed herbicide.

When the tree roots take up too much of the herbicide diuron, major leaf veins turn yellowish.

Leaves deficient in zinc develop chlorotic areas between the veins.

When magnesium is deficient, yellowing of leaf tissue begins at tip and margins and moves inward. A somewhat V-shaped, dark green area usually remains at the leaf base.

Manganese deficiency causes a fine mottling of mostly young leaves, with pale green to yellowish interveinal areas and dark green, sharply contoured veins.

If a growth regulator, such as 2,4-D, is applied incorrectly, leaves cup or curl and, in severe cases, become bleached.

SYMPTOMS AFFECTING GROWTH HABIT AND YIELD

Three quite serious citrus diseases affect overall growth habit and yield, often without producing distinctive symptoms on the trunk, roots, or leaves. Stubborn disease is caused by a mycoplasma, tristeza is caused by a virus, and lemon sieve tube necrosis is an inherited disorder of lemon trees.

Stubborn Disease
Spiroplasma citri

Stubborn disease is endemic in the warm inland growing areas, where it affects primarily sweet orange, grapefruit, and tangelo trees. The disease is more a problem in young orchards than in mature groves. The pathogen is a mycoplasma, which spreads by leafhopper feeding and by grafting and budding. In severe cases, stubborn disease causes a dramatic yield loss.

Symptoms and Damage

Stubborn disease is often difficult to diagnose, especially in the early stages of disease development when symptoms are subtle or when other disorders are present. Symptoms also vary somewhat among cultivars.

The most obvious symptoms of stubborn-infected trees are a low yield of abnormally small fruit, the absence of fruit, and the stunted, feathery growth of the canopy. The leaves are small and grow upright close to the stems. The trees usually develop unseasonable growth flushes and blossoms. The few fruit produced remain small and are lopsided. The best way to see the off-centered navel and uneven sides is to cut a fruit in half.

Certain other fruit symptoms may appear. Depending on the ripening stage of the fruit, you may see stylar end greening; the blossom end of the fruit remains green while the stem end becomes colored. Fruit of seedy cultivars have dark colored, small seeds aborted early in their development. The fruit may have an insipid or bitter flavor; on some cultivars, they also become acorn shaped.

If young trees are infected, the entire tree remains small and unproductive. If mature trees become infected, a single branch may show symptoms, and the disease may or may not spread slowly throughout the tree. Stubborn does not kill trees but stunts growth and inhibits fruit production.

Seasonal Development

Because hot, dry weather favors the development and spread of the stubborn pathogen, it has become a problem in the San Joaquin and desert valleys. Contaminated budwood and leafhoppers are the two sources of infection. At least two leafhopper species, the beet leafhopper, *Circulifer tenellus*, and *Scaphytopius nidridus*, have been shown to carry the pathogen and to transmit it during feeding on citrus.

Management Guidelines

Management of stubborn disease focuses on preventing the disease and avoiding its spread. Preventive measures mainly apply to nursery practices, such as maintaining stubborn-free mother trees for budwood. Grafting budwood onto indicator seedlings or culturing leaf and fruit samples in the lab can determine the presence of the stubborn organism. No commercial laboratories, however, are currently equipped to carry out these tests.

When planting an orchard, obtain trees from an area that does not have a high incidence of stubborn disease. In an established orchard, observe the trees carefully for any signs of stubborn disease in late fall or early winter. A sparse crop—a useful diagnostic symptom—becomes apparent as fruit color changes to orange. Map or flag the trees suspected of being infected and recheck the orchard several times during the year to confirm your diagnosis. Replace diseased and unproductive trees. Topworking is not advisable because the pathogen moves freely between the scion and rootstock.

Treatment of leafhoppers in the field does not prevent the spread of the mycoplasma. Studies are underway to develop methods to protect young trees in nurseries from the insect vector.

Tristeza Disease Complex
Tristeza virus

Tristeza is widespread throughout southern California, but so far the concerted efforts of various groups have prevented tristeza from spreading into the San Joaquin and desert valleys.

Tristeza diseases, including tristeza, quick decline, seedling yellows, and stem pitting, are different syndromes caused by isolates of the tristeza virus. They differ in their virulence and their reaction to different host cultivars. Tristeza, or field tristeza, is the most common disease that results in the rapid or "quick decline" symptoms when sweet orange is grafted on sour orange rootstock. This isolate can also cause a slow decline on certain other rootstocks. For example, sweet orange on Troyer citrange can develop a slight decline, from which the trees almost fully recover with time. Grapefruit is somewhat susceptible, and affected trees are in a state of chronic decline.

The more virulent isolate of the tristeza virus, seedling yellows, has been found in certain locations in southern California. The name refers to symptoms that appear on seedlings grafted with infected budwood. This disease is of concern because all commonly used rootstocks are susceptible and the melon aphid readily transmits the seedling yellows isolate.

Other isolates can cause stem pitting on sweet orange regardless of the rootstocks. Stem pitting, however, is not a reliable symptom for filed diagnosis.

Symptoms and Damage

Susceptible rootstock/scion combinations infected with the virus show symptoms similar to those caused by other diseases that injure roots or girdle the crown. Trees infected with tristeza show light green foliage, poor growth flushes, and some leaf drop. The trees may produce a heavy crop of small fruit because the girdling at the bud union prevents starch transport to the roots. Diseased young trees bloom early and abundantly and begin producing fruit 1 to 2 years before healthy ones do.

By the time the aboveground symptoms appear, many feeder roots have died. The girdling of the food-conducting tissue just below the bud union results in starvation of the roots, and feeder roots die from the periphery inward. Depending on the extent of girdling, part or all of the feeder root system may be destroyed, corresponding

Young trees infected with stubborn disease are stunted and produce only few small fruit.

to various stages of decline of the scion. Infected trees on Troyer usually reach an equilibrium in their decline but never regain full productivity.

Seedling yellows cannot be identified in the field. The symptoms do not differ from those of tristeza, and positive identification requires an indexing text.

Seasonal Development

Tristeza virus is spread through budding and grafting or by aphids feeding on citrus. The melon aphid, *Aphis gossipyii*, is the vector for all tristeza isolates found in California; it does not, however, transmit all isolates equally well. Symptoms of tristeza become more apparent during the hot summer months, when increased water needs cannot be met by the declining root system. Changes in virus virulence may occur over time and affect the current tolerance of many cultivars. Tristeza, seedling yellows, and stem pitting may occur together in the same tree.

Management Guidelines

Management of the tristeza complex depends largely on preventive measures, such as using tolerant rootstocks and tristeza-free propagation material. However, because of the insect vector, disease spread cannot be prevented completely.

Many rootstock/scion combinations are tolerant to tristeza. Sweet orange on sour orange, Sampson tangelo, shaddocks, and possibly grapefruit on sour orange are susceptible. Troyer citrange and *Citrus macrophylla* rootstocks are affected to a lesser extent by the disease in Ventura County. *C. macrophylla* is also susceptible to stem pitting. All rootstocks are susceptible to infection by the seedling yellows virus. When choosing a rootstock, also consider your local conditions concerning soil, climate, and other disease problems (Table 2).

The few fruit produced by trees infected with stubborn remain small and are lopsided. Here a fruit is cut in half to show the off-centered fruit axis.

If a sweet orange scion on a sour orange rootstock is infected with tristeza virus, the tree declines rapidly, with leaves and fruit drying on the tree. On other rootstocks, trees decline slowly and often recover.

When grafting or topworking, use only certified, virus-free budwood. The Citrus Clonal Protection Program (CCPP) provides virus-free and true-to-type bud lines to nurserymen and growers in California. Contact your county agricultural commissioner's office for listings of nursery services participating in the CCPP program.

Observe quarantine restrictions to avoid spreading tristeza. No plants or plant parts should be shipped from infected southern California districts to areas where tristeza is not present or is localized, such as the San Joaquin Valley or Coachella Valley.

Support eradication efforts in designated districts. The Central Valley Tristeza Control Agency, the California Department of Food and Agriculture, and county agricultural commissioners of Fresno, Tulare, and Kern County work together to keep the central valley largely free of tristeza. The agricultural commissioner's office at Riverside conducts a similar program for the Coachella Valley. Agency personnel carry out surveys, perform indexing routinely, and remove infected trees.

In southern California, where tristeza is widespread, you may want to remove infected trees only when they become unproductive. At the University of California at Riverside, indexing efforts have recently been stepped up to eradicate the more severe seedling yellows and stem pitting isolates from the experiment station. You can help in eradicating other localized infections of seedling yellows by removing old-line Meyer lemon, a carrier of seedling yellows virus, and replacing it with the new, virus-free Meyer lemon (CV 1319 and 333).

Lemon Sieve Tube Necrosis

Lemon sieve tube necrosis (LSN) is a common, inherited disorder of lemon trees in coastal areas. Eureka budlines are more severely affected than Lisbon lemons. In intermediate areas, the disease is less severe, and in the San Joaquin and Coachella valleys, LSN does not result in noticeable decline.

Trees with the disease go through a cyclic decline. About 4 or 5 years after planting, the older food-conducting sieve tubes in the phloem near the bud union die. Several years later, younger sieve tubes also die, severely restricting food transport to the roots. Many feeder roots starve, fruit ripen prematurely, shoots grow poorly, and some leaves turn yellow and drop. The dieback stimulates new cambium and phloem production, and the tree recovers temporarily. Once the new sieve tubes also become necrotic, the decline process starts again. Only a microscopic analysis can reveal the collapsed sieve tubes.

Only certain budlines are affected by this inherited disorder. Before planting Eureka lemons, obtain the most recent recommendations from your farm advisor.

Nematodes

Nematodes are small worms that live in soil, water, and plant tissues. Some damage plants, others feed on fungi, bacteria, or other nematodes. The major species attacking citrus is the citrus nematode, *Tylenchulus semipenetrans*, which is present in about 50 to 60% of mature citrus orchards and can reproduce in all soil types. The host range includes citrus, grape, lilac, olive, and persimmon. The less important sheath nematode, *Hemicycliophora arenaria*, occurs on citrus in the desert. It has a broad host range, including many native plants and row crops.

Description and Biology

The citrus nematode, *Tylenchulus semipenetrans*, is so small that you will need a microscope to see it clearly. Identification of nematodes is difficult, so consult a

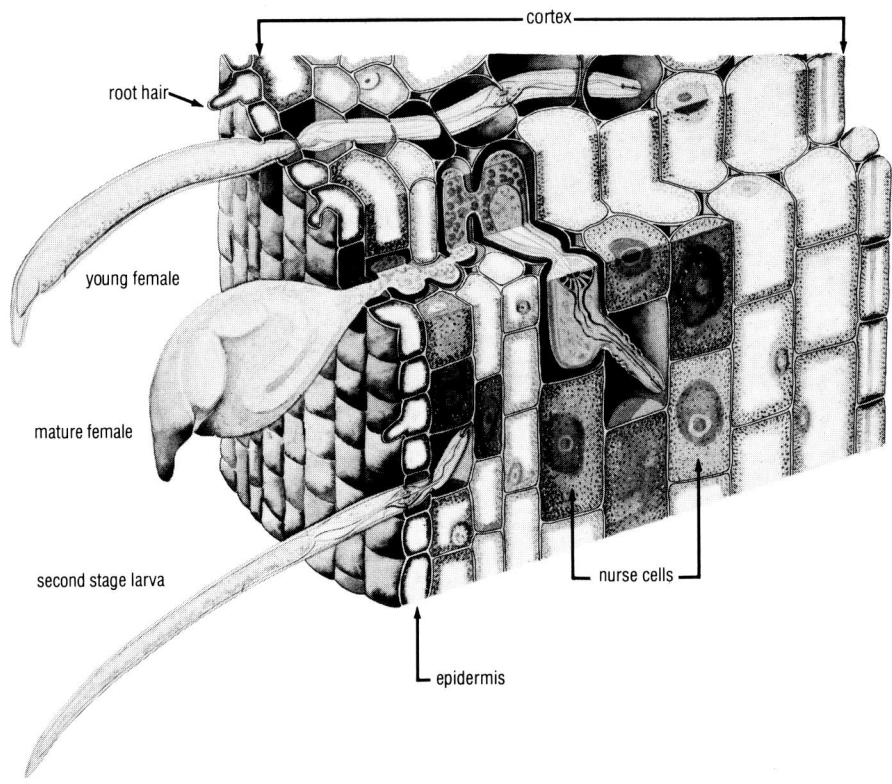

Figure 62. Citrus nematode feeding on a citrus root. The second stage larva and young female enter the root and feed in different cell layers of the cortex. The mature female settles with her front end in the permanent feeding site within the nurse cells, while her swollen rear end remains outside the root.

nematologist to identify the species or verify the presence of citrus nematodes.

Female citrus nematodes spend most of their life cycle partially burrowed in the feeder roots of the citrus tree. Mature females lay their eggs into a gelatinous material that covers their posterior. Young larvae hatch while still enclosed in the egg sac. After molting once they move freely in the soil. The second instar females feed on the outside of rootlets for a while and then burrow into the feeder roots. Less than 10% of the larvae in the soil reach and infect the roots. Females molt three more times and then settle with their heads deep inside the root cortex, with their rear end outside in the soil (Figure 62). The posterior portion enlarges as the reproductive organs develop and is covered by the gelatinous material during reproduction.

The female life cycle from egg to egg takes 6 to 8 weeks under optimum temperatures of 25 to 30°C (77 to 86°F). Male nematodes live only about a week after leaving the egg; they do not feed and do not burrow into roots. Citrus nematode larvae are not active when soil temperatures drop below 16°C (60°F). Eggs and first instar larvae can survive inactive in the gelatinous egg masses for more than a year in old roots.

Damage

Rootstock susceptibility, nematode population density, and the age and health of trees largely determine the damage caused by a citrus nematode infestation. Susceptible trees planted in lightly infested soil may grow for many years without apparent damage and then decline slowly. Resistant rootstocks generally do well even in heavily infested soils. If, however, a heavily infested orchard site is replanted with a susceptible rootstock without soil fumigation, the roots of the young trees will soon be heavily parasitized, tree growth will be stunted, and fruit production reduced. This condition is also referred to as the *citrus replant problem*.

Young citrus trees suffer more from a nematode infestation than mature trees, which can tolerate many of these small root parasites. Adverse effects become apparent when trees are already weakened by other stress factors, such as Phytophthora root rot or water excess.

The aboveground symptoms of a nematode infestation are difficult to diagnose. Reduced fruit size and yield or, in severe cases, twig dieback, and small leaves and fruit may be caused by nematodes but also symptoms caused by soil pathogens or adverse soil conditions. The belowground symptoms are similar to those caused by a Phytophthora infection: "dirty" appearance of the feeder roots and, in severe infestations, the lack of feeder roots. Often Phytophthora root rot and citrus nematodes occur together in roots, and both may contribute to the damage symptoms.

Guidelines for Managing Nematodes

Monitoring. Before planting or replanting a citrus orchard, obtain a professional soil analysis; the analysis will help you determine the potential for nematode damage and plan a management strategy. In an established orchard, a soil analysis will confirm visible symptoms that may be present (see below). Some laboratories collect samples, or you may have to do it yourself. For a list of professional laboratories, see UC Agricultural Sciences Publication 3024, *Commercial Analytical Laboratories in California for Agricultural Testing*, or consult your farm advisor.

To collect samples before planting, visually divide the orchard into sampling blocks representing differences in soil texture, drainage pattern, or cropping history. In an established orchard irrigated by sprinklers or furrows, collect soil and root samples at the drip line of trees that show symptoms and samples from adjacent, healthy looking trees for comparison. In drip-irrigated orchards, take samples around emitters where feeder roots are most abundant. The soil should not be too dry or too wet. You can sample fallow land at any time of the year. The best time to sample an established orchard is March through April, so that measures can be taken, if necessary, to protect the spring growth flush of the roots. In loamy soils, sampling down to 60 cm (24 inches) is sufficient; in sandy soils, take samples to a depth of 90 cm (36 inches). Use a soil auger, Viehmeyer tube, or shovel (Figure 16, p. 22). A soil auger (7.5 cm or 3 inches in diameter) is convenient for depths to 60 cm in sandy soils. To sample deeper than 60 cm, a Viehmeyer tube is recommended to reduce the soil volume taken. The tube can easily be hammered down to 1.2 m (48 inches); however, the amount of roots collected will be much smaller than with a soil auger.

From each sampling block, collect 10 to 20 cores or subsamples. Combine the subsamples, mix thoroughly, and pour the soil and roots into durable plastic bags or other moisture-proof containers. Seal tightly and place bags in the shade until you have taken the last sample. Attach labels providing name and address, location of the orchard, sample block, soil texture, cropping history, notable symptoms and, if possible, rootstock and soil and air temperature; this information is critical for a meaningful analysis. Send or deliver the samples to the lab as soon as possible. Ship them in a cardboard box insulated with newspaper, or in a styrofoam ice chest. If any delay occurs, keep the samples in a cool place (5° to 10°C or 41° to 50°F).

Most labs extract nematode larvae from soil samples using the Baermann funnel or the elutriation/flotation method. The method used and often the extraction efficiency is reported together with the results. Larval counts are generally sufficient for estimating relative infestation levels (see below). Extracting females from the citrus roots, however, is more accurate, especially when checking the success of a chemical treatment at the end of the season when larval counts are usually low because of low temperatures. Commercial labs currently do not analyze citrus roots for female nematodes. For more information on extraction procedures, contact the Department of Nematology, University of California at Riverside.

Interpreting Soil Analysis. Although it will vary greatly with soil moisture, soil type, and temperature, the number of nematodes in the soil, as determined by soil analysis, can give some indication about the damage potential of an infestation. Samples cannot provide an accurate prediction of yield at the end of the season since many other factors, including alternate bearing habit of citrus and other pest problems, may influence yield. Table 10 shows the average number of larvae and females at different sampling times; different soil types are not taken into account. The table gives a rough estimate of low, medium, and high populations. A preplant treatment is recommended at all levels when replanting an orchard with either a tolerant or a susceptible rootstock. At low levels in an established orchard, a treatment is not economical, but you should continue sampling at least once a year to see if the population remains low. At medium levels, treatment may be advantageous if the site has a history of nematode damage. At high levels, a treatment can prevent substantial reduction in fruit size and yield. In both cases successful treatment requires precise and repeated applications. Available postplant nematicides are expensive; you have to weigh treatment costs and age and condition of the orchard as well as projected crop loss.

Table 10. Rating of Population Levels of the Citrus Nematode Larvae and Females as Determined by Soil Analysis[1]

Population level	LARVAE per 500 g soil		FEMALES per 1 g roots	
	Feb.-Apr.	May-July	Feb.-Apr.	May-June
low	<2000	<4000	<100	<300
medium	>5000	>8000	>400	>700
high	>12,000	>18,000	>1100	>1400

1. Samples taken at 6 cm (0-2) ft. depth with Viehmeyer tube; extraction with Baermann funnel; nematode numbers adjusted to 100% extraction efficiency; < = less than, > = greater than.

The number of females per unit of feeder roots is more representative of the damage potential to the tree than the number of free larvae in the soil. If the population of females exceeds the medium level (Table 10), tree growth and fruit production are likely to be reduced.

Prevention and Control. Before planting a citrus orchard, consider the cropping history of the site and the drainage from surrounding orchards. If the site has recently been cleared of citrus or other host plants, or if runoff from an infested orchard has entered the property, the soil is likely to be infested with the citrus nematode; it is a poor site for a new citrus planting unless a resistant rootstock is used or other management actions are taken.

Choosing a resistant or tolerant rootstock is a good strategy, whether or not the soil is infested with nematodes. Trifoliate orange is resistant to the citrus nematode; Troyer citrange is tolerant but may eventually become infected as new biotypes of the citrus nematode develop (Table 2, p. 17). Where a resistant rootstock cannot be used, rotation to an annual crop for 1 to 3 years and subsequent fumigation can reduce the population of the citrus nematode so that young trees are better able to establish themselves. The rotation allows the decay of woody roots that harbor nematodes. For the most effective fumigation, carefully observe labeling instructions concerning land preparation, soil type, temperature, wind, residual crop roots, dosages, and waiting periods between treatment and planting.

Good sanitation practices are important in preventing a nematode infestation. Buy your trees from a nursery that sells nematode-free material. Clean equipment thoroughly when moving into another orchard. Do not irrigate with runoff water from areas infested with nematodes.

Little is known about how to use cultivation and irrigation practices to prevent a buildup of nematode populations in an orchard. In general, any practices that favor root growth and reduce tree stress help trees cope better with nematodes. Nematode-trapping and parasitic fungi, predaceous nematodes, and soil mites, may play a significant role in reducing populations of the citrus nematode. Such natural enemies build up in many soils over time and are often found in substantial numbers along the surface of citrus feeder roots, suggesting that citrus nematodes might be more damaging if they were not present. In general, loam soils high in organic matter have a greater diversity of natural enemies and high populations of individual species than sandy soils low in organic matter. Nematode-trapping fungi especially are remarkably widespread in mature citrus orchard soils. Nematicides, particularly preplant fumigants, largely eliminate these beneficial organisms.

Once you detect nematode damage in a citrus orchard, control options are limited. Postplant nematicides are available but are expensive and require precise and repeated application to be effective; you have to weigh projected loss against age of the orchard, infestation level, and treatment cost. For pesticide recommendations, see the latest University of California *Treatment Guide for California Citrus Crops.*

Weeds

Plants growing on the orchard floor become weeds when they compete with citrus trees for water, nutrients, and light, contribute to pest problems, interfere with cultural operations, or increase frost hazard. Competition may vary with weed species and density but is particularly strong in newly planted orchards; stressed young trees grow slower and may become less tolerant of damage from insects and diseases. Weeds around tree trunks may create a favorable environment for pathogens that infect the trunk and roots. As the trees grow older, the tree skirts and canopies shade part of the orchard floor and reduce weed growth.

Most weed species found in citrus orchards are either annual or herbaceous perennial plants. Annual plants complete their cycle in one growing season. Winter annuals germinate in the fall, grow during the winter, and flower and produce seed in the spring before dying in early summer. Common winter annuals found in citrus groves are annual bluegrass, burning nettle, chickweeds, common groundsel, foxtails, henbit, miners lettuce, fiddleneck, filaree, little mallow (cheeseweed), mustards, shepherdspurse, and wild barley. Summer annuals germinate in the spring or early summer and flower and produce seed in the fall before dying in the winter. Major species include barnyardgrass (watergrass), crabgrasses, common lambsquarters, flax-leaved fleabane, horseweed, lovegrass, pigweeds, puncturevine, spotted spurge, purslane, sprangletop, nightshades, turkey mullein, vinegar weed, and witchgrass.

In California's mild climate, certain annuals may behave as biennials or short-lived perennials—for example, horseweed, little mallow, and sweet clovers. Biennials complete their life cycle in two growing seasons, producing vegetative parts in the first growing season and flowers and seeds in the second. Perennials, which can live 3 years or longer, may be either herbaceous or woody plants. Some herbaceous perennials die back during the winter but regrow during the spring or early summer from underground rhizomes, bulbs, tubers, or crowns on tap roots. Commonly found herbaceous perennials in citrus groves include bermudagrass, dallisgrass, field bindweed, johnsongrass, and nutsedges. Because of their underground resources, perennials are much more difficult to control than annual weeds.

Spotted spurge provides a good ground cover because it is low growing and is not a strong competitor for nutrients and water.

Strip control is practiced in some citrus orchards. A strip along the trees is kept free of weeds with herbicides, whereas the middle of the row is mowed.

Guidelines for Managing Weeds

Prevention

Careful management practices in and around the orchard help limit weed infestations. To prevent the spreading of weeds, make sure irrigation ditches are free of weeds. If necessary, install screens to keep out dislodged weeds and weed seeds. Water management is particularly important. Provide good drainage; high moisture in areas such as furrow bottoms, at furrow ends, and around stand pipes greatly favors weed growth. Where furrow irrigation is used on slow draining soil, change to shorter furrows or establish lateral furrows halfway into the tree rows to reduce the time water stands in the furrows. You can discourage the establishment of seedlings by letting the top 2 or 3 inches of soil dry completely between furrow or sprinkler irrigations.

Biological control of weeds by plant pathogens and insects is often at work in orchards and provides some degree of control. For example, the puncture vine seed and stem weevils help reduce the number of puncture vines in orchards. Generally, however, the biological agents known today are not effective enough to provide substantial control of weeds in citrus orchards.

Control Methods

In most citrus-growing areas, groves are maintained fairly weed free with herbicides. Herbicides provide a convenient and efficient control of most weeds, facilitate irrigation and other cultural operations in the grove, and create a bare orchard with less frost hazard during the winter.

Certain problems are associated with total reliance on herbicides. In orchards planted on slopes, soil erosion can be substantial in bare orchards. On certain soils, compaction and the development of a silty surface layer that impedes water penetration may be a problem. Repeated, shallow cultivation or the application of a mulch may be needed as a remedy. If a particular herbicide is used repeatedly, species that it does not kill may thrive in the absence of competition from other species and may become the dominant weeds in the orchard. For example, simazine eliminates most broadleaf annuals, so that grasses, such as barnyardgrass, crabgrass, and witchgrass, build up and require additional postemergence management. Spotted spurge and some biotypes of groundsel also survive applications of simazine, so they may be common in treated orchards. Common groundsel, speedwell, spotted spurge, and turkey mullein can become dominant in orchards treated with diuron. Preemergence herbicides may favor established perennial weeds, which are not controlled by most preemergence materials and spread rapidly in the absence of annuals.

A ground cover is maintained in some citrus groves, mainly in the northern California growing regions and on hilly terrain. A ground cover of volunteer weeds or a sown cover crop prevents soil erosion and improves water penetration and soil structure on certain soil types. Spotted spurge can provide a good cover crop. It is low growing, does not need mowing, and is not a strong competitor for nutrients and water. It is resistant to commonly used preemergence herbicides. Other ground covers can be managed by complete mowing or by mowing the row middles while keeping a strip along the tree rows free of weeds with herbicides. Repeated mowing favors the establishment of perennials, particularly bermudagrass; such perennials are deep rooted and more competitive with citrus than annuals like spotted spurge. In many areas, the additional water requirements of maintaining a ground cover make it impractical.

Weeds are rarely managed with regular cultivation because of its many disadvantages. Tillage destroys the valuable layer of feeder roots which absorbs nutrients, water, and oxygen in the top soil. The injury to the root system also provides entry sites for disease organisms. Discing contributes to soil erosion, especially on sloping land, and to soil compaction. If the soil is dry, cultivating creates dust, which interferes with the biological control of insect and mite pests. Discing may also increase the weed population by bringing buried seeds to the surface or spreading rhizomes, tubers, or stolons throughout the orchard.

Use of Herbicides. Herbicides are applied either to the soil or to the weed foliage. Preemergence (soil-applied, residual) herbicides are sprayed onto the soil just before an irrigation or rainfall, so that the water carries the chemical into the soil, where the weed seeds germinate. Examples are simazine, bromacil, diuron, and a bromacil/diuron mixture. These herbicides are most effective on germinating weeds; at rates used in citrus, they generally do not control established weeds. Diuron and bromacil, however, are also foliar active on certain young, established weeds if additional surfactant is added to the spray solution. Preemergence herbicides can provide control for up to a year, depending on the solubility of the material, soil properties, frequency and method of irrigation, weed species, and dosage applied. Leaching from the soil is more extensive on sandy than on clay soils. Prolonged moist conditions during heavy winter rains, in furrow bottoms, or around low-volume emitters during irrigation favor breakdown and leaching of herbicides. Bromacil leaches more rapidly than simazine, diuron, and napropamide and is thus less effective under low-volume irrigation. Splitting a treatment into two or more applications can prolong the control provided by the herbicides.

Postemergence (foliar-applied) herbicides are used on established weeds. They act either by contact or by translocation throughout the plant. Contact herbicides kill

	PREEMERGENCE						POSTEMERGENCE				
	SIMAZINE (PRINCEPS®)	DIURON (KARMEX®)	BROMACIL (HYVAR X®)	DIURON/BROMACIL (KROVAR I®)	TRIFLURALIN (TREFLAN®)	NAPROPAMIDE DEVRINOL®)	DINOSEB* (VARIOUS NAMES)	PARAQUAT* (PARAQUAT)	GLYPHOSATE (ROUNDUP®)	MSMA (BUENO 6®)	WEED OIL
ANNUAL BROADLEAVES											
Burning nettle	controlled	controlled	controlled	controlled	partially controlled	partially controlled	controlled	controlled	controlled	not controlled	controlled
Cheeseweed (Malva)	controlled	controlled	controlled	controlled	partially controlled	partially controlled	controlled	not controlled	partially controlled	not controlled	partially controlled
Chickweed	controlled	controlled	controlled	controlled	controlled	controlled	controlled	controlled	controlled	controlled	controlled
Cocklebur	controlled	controlled	controlled	controlled	not controlled	controlled	partially controlled	controlled	controlled	partially controlled	controlled
Common groundsel	controlled	not controlled	controlled	controlled	not controlled	controlled	partially controlled	controlled	controlled	not controlled	controlled
Cudweed	controlled	controlled	controlled	controlled	not controlled	controlled	partially controlled	not controlled	controlled	not controlled	controlled
Clovers	controlled	controlled	partially controlled	controlled	not controlled	partially controlled	controlled	partially controlled	partially controlled	not controlled	controlled
Fiddleneck	controlled	controlled	controlled	controlled	controlled	controlled	controlled	controlled	controlled	not controlled	controlled
Filaree	not controlled	controlled	controlled	controlled	controlled	controlled	controlled	controlled	partially controlled	not controlled	controlled
Flax-leaved fleabane	controlled	controlled	controlled	controlled	not controlled	not controlled	controlled	controlled	controlled	not controlled	controlled
Groundcherry	controlled	controlled	controlled	controlled	not controlled	not controlled	controlled	controlled	controlled	partially controlled	controlled
Henbit	controlled	controlled	controlled	controlled	partially controlled	partially controlled	controlled	controlled	controlled	not controlled	controlled
Horseweed	controlled	controlled	controlled	controlled	not controlled	not controlled	controlled	controlled	controlled	not controlled	controlled
Knotweed	controlled	controlled	controlled	controlled	controlled	controlled	partially controlled	partially controlled	controlled	not controlled	controlled
Lambsquarters	controlled	controlled	controlled	controlled	controlled	controlled	controlled	partially controlled	controlled	not controlled	controlled
London rocket	controlled	controlled	controlled	controlled	not controlled	partially controlled	controlled	controlled	controlled	not controlled	controlled
Mustards	controlled	controlled	controlled	controlled	not controlled	partially controlled	controlled	controlled	controlled	not controlled	controlled
Nightshades	controlled	controlled	controlled	controlled	not controlled	not controlled	controlled	controlled	controlled	partially controlled	controlled
Pigweeds	controlled	controlled	controlled	controlled	controlled	controlled	controlled	controlled	controlled	not controlled	controlled
Pineappleweed	controlled	controlled	controlled	controlled	not controlled	controlled	controlled	controlled	controlled	not controlled	controlled
Popcorn flower	controlled	controlled	controlled	controlled	controlled	not controlled	controlled	controlled	controlled	not controlled	controlled
Prickly lettuce	controlled	controlled	controlled	controlled	not controlled	controlled	controlled	controlled	controlled	not controlled	controlled
Puncturevine	partially controlled	partially controlled	controlled	controlled	controlled	partially controlled	controlled	controlled	controlled	partially controlled	controlled
Purslane	controlled	controlled	controlled	controlled	controlled	controlled	controlled	controlled	controlled	not controlled	controlled
Red maids	controlled	controlled	controlled	controlled	controlled	not controlled	controlled	controlled	controlled	not controlled	controlled
Russian thistle	controlled	not controlled	controlled	controlled	partially controlled	partially controlled	controlled	controlled	controlled	not controlled	controlled
Shepherdspurse	controlled	controlled	controlled	controlled	not controlled	partially controlled	controlled	controlled	controlled	not controlled	controlled
Sowthistle	controlled	controlled	controlled	controlled	not controlled	controlled	controlled	controlled	controlled	not controlled	controlled
Speedwell	controlled	controlled	controlled	controlled	partially controlled	controlled	controlled	controlled	controlled	not controlled	controlled
Spotted spurge	not controlled	not controlled	partially controlled	partially controlled	not controlled	not controlled	partially controlled	controlled	controlled	not controlled	controlled
Turkey mullein	not controlled	not controlled	partially controlled	partially controlled	not controlled	not controlled	controlled	partially controlled	controlled	not controlled	controlled
Wild radish	controlled	controlled	controlled	controlled	not controlled	partially controlled	partially controlled	partially controlled	controlled	not controlled	controlled
ANNUAL GRASSES											
Annual bluegrass	partially controlled	controlled	controlled	controlled	controlled	controlled	not controlled	controlled	controlled	controlled	controlled
Annual ryegrass	partially controlled	controlled	controlled	controlled	controlled	controlled	controlled	controlled	controlled	partially controlled	partially controlled
Barnyardgrass	partially controlled	controlled	partially controlled	controlled	controlled	controlled	not controlled	partially controlled	controlled	partially controlled	controlled
Bromegrass	not controlled	controlled	controlled	controlled	controlled	controlled	not controlled	controlled	controlled	partially controlled	controlled
Canarygrass	partially controlled	controlled	controlled	controlled	controlled	controlled	not controlled	controlled	controlled	not controlled	controlled
Crabgrass	not controlled	controlled	partially controlled	controlled	controlled	controlled	not controlled	controlled	controlled	controlled	controlled
Foxtail barley	partially controlled	controlled	controlled	controlled	controlled	controlled	not controlled	controlled	controlled	not controlled	controlled
Johnsongrass (seedl.)	partially controlled	controlled	controlled	controlled	controlled	controlled	not controlled	controlled	controlled	controlled	controlled
Lovegrass	partially controlled	controlled	partially controlled	controlled	controlled	controlled	not controlled	controlled	controlled	partially controlled	controlled
Sandbur	not controlled	controlled	partially controlled	controlled	controlled	controlled	not controlled	partially controlled	controlled	not controlled	controlled
Wild oats	partially controlled	partially controlled	controlled	controlled	partially controlled	controlled	not controlled	controlled	controlled	not controlled	controlled
Witchgrass	not controlled	controlled	controlled	controlled	controlled	controlled	not controlled	controlled	controlled	ND	partially controlled
Sprangletop	not controlled	partially controlled	controlled	controlled	controlled	ND	not controlled	controlled	controlled	not controlled	partially controlled
PERENNIALS											
Bermudagrass	not controlled	not controlled	controlled	partially controlled	partially controlled	not controlled	not controlled	partially controlled	controlled	not controlled	partially controlled
Dallisgrass	not controlled	not controlled	partially controlled	partially controlled	not controlled	not controlled	not controlled	not controlled	controlled	partially controlled	partially controlled
Field bindweed	not controlled	not controlled	partially controlled	not controlled	controlled	not controlled	not controlled	not controlled	controlled	partially controlled	partially controlled
Johnsongrass	not controlled	not controlled	not controlled	not controlled	partially controlled	not controlled	not controlled	partially controlled	controlled	controlled	partially controlled
Nutsedges	not controlled	not controlled	controlled	partially controlled	not controlled	partially controlled	not controlled	not controlled	partially controlled	partially controlled	partially controlled

controlled — partially controlled — not controlled — ND = no data

* Permit required

those parts of the plant that are actually sprayed, making good coverage and wetting essential. Examples are weed oils, used alone or fortified with dinoseb, and water solutions of paraquat or emulsions of water, weed oil, and dinoseb. A single spray kills susceptible annual weeds; retreatment is necessary if regenerating perennials are present or if annual weeds reestablish themselves from seeds. Contact herbicides are most effective when applied to weed seedlings or young weeds. When treating at this stage, wetting plant surfaces is easier and less material is needed for effective control than on old, hardened growth. Translocated or systemic herbicides do not require thorough coverage because the material is transported from the sprayed part to other plant parts, such as roots and rhizomes. Foliar-applied, translocated materials, such as glyphosate and MSMA, are therefore more effective in killing perennials.

Choose herbicides and application rates according to weed species, soil type, irrigation method, and age of trees. No single material registered for citrus will control all weed species; combinations or sequential applications of different herbicides may provide better control than one compound alone in many situations. A given dosage of preemergence herbicide is more toxic in sandy soils or soils low in organic matter than an equal dosage in soils high in clay or organic matter. Certain herbicides may injure young citrus trees. Bromacil may leach in irrigation or rain water and injure neighboring crops. Follow all label precautions carefully. Use the lowest effective rate to avoid illegal residues on or in the fruit. Carefully calibrate and check the functioning of spray equipment. The procedures are described in University of California Agricultural Sciences Publications Leaflet 2710, *Calibration of Herbicide Sprayers*). Figure 63 shows the approximate control provided by herbicides currently registered for citrus. Details about registration status and the use of individual herbicides are given in University of California Agricultural Sciences Publications Leaflet 2979, *Weed Management Guide for Citrus*, and the latest *Treatment Guide for California Citrus Crops*, Leaflet 2903.

Citrus is more tolerant than deciduous trees to some herbicides, but injury can occur from incorrect use. Injury symptoms vary in intensity and appearance, depending on dosage, method of application, and soil type.

Monitoring and Control Programs

To plan a weed management program, you must know the weed species present, and their abundance and location in the orchard. Conduct a survey at least twice a year, in late winter and summer, and keep records of your observations. Pay special attention to perennials, and check fencerows and ditchbanks. A map can be helpful in locating trouble spots infested with perennials or resistant species, moist areas favoring weed growth, or sources of reinfestation from surrounding land. Record your monitoring results in a form such as the one given in Figure 64, or make your own form. You can use a scale from 1 to 5 to indicate the level of infestation: 1 = very few weeds; 2 = light infestation; 3 = moderate infestation; 4 = heavy infestation; 5 = very heavy infestation. Monitoring information collected over several years is invaluable in determining changes in the weed flora and adjusting the control program.

ORCHARD _____

MONITORING DATE _____

HERBICIDE _____

APPLICATION DATE _____

COMMENTS _____

PERENNIALS
- ☐ Bermudagrass
- ☐ Dallisgrass
- ☐ Field bindweed
- ☐ Johnsongrass
- ☐ Nutsedges
- ☐ _____

ANNUAL GRASSES
- ☐ Annual bluegrass
- ☐ Barnyardgrass
- ☐ Bearded sprangletop
- ☐ Crabgrass
- ☐ Foxtails
- ☐ Lovegrass
- ☐ Witchgrass
- ☐ _____
- ☐ _____

ANNUAL BROADLEAF WEEDS
- ☐ Burning nettle
- ☐ Cheeseweed (Malva)
- ☐ Common groundsel
- ☐ Common lambsquarter
- ☐ Fiddleneck
- ☐ Filaree
- ☐ Flax-leaved fleabane
- ☐ Horseweed
- ☐ Mustards
- ☐ Nightshades
- ☐ Pigweeds
- ☐ Puncturevine
- ☐ Spotted spurge
- ☐ Turkey mullein
- ☐ _____
- ☐ _____
- ☐ _____

Figure 64. Example of a weed monitoring sheet. You can use a scale from 1 to 5 to indicate the level of infestation: 1 = very few weeds; 2 = light infestation; 3 = moderate infestation; 4 = heavy infestation; 5 = very heavy infestation.

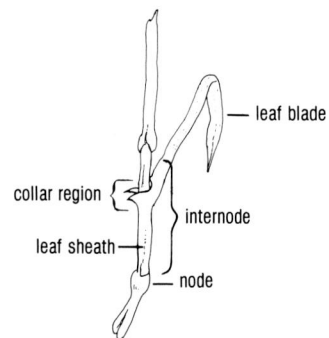
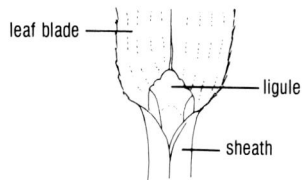
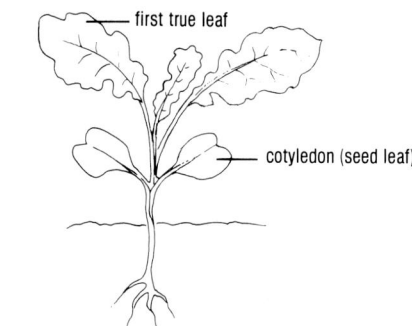
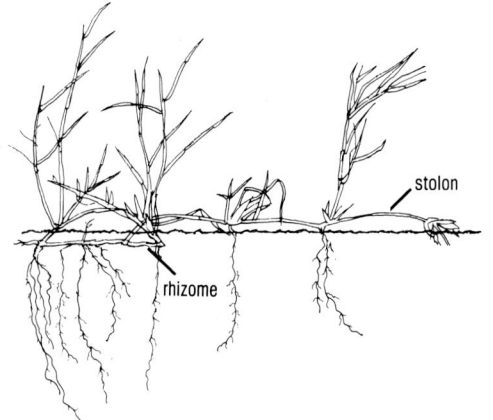

Figure 65. Vegetative parts commonly used for identifying weeds.

Management is easier if you identify and control weeds in the seedling stage. The photos and description in this section should help identify some of the most common weeds in citrus orchards. A few necessary terms are illustrated in Figure 65 and explained in the glossary.

There are three categories of weeds: broadleaves, grasses, and sedges. Broadleaf weeds at emergence have two seed leaves, often of characteristic shape, color, and texture. Seed leaves and usually the first true leaves differ from later leaves. Grass seedlings are difficult to identify, but the collar region is generally the most distinguishing feature. Perennial grasses may also have characteristic rhizomes or stolons. Sedges are grasslike but typically have triangular stems. In most instances, a combination of characteristics will identify a species, so examine several young plants to confirm your identification.

New Orchard Sites and Young Plantings. Weed control starts before planting an orchard. Survey the site to determine which weed species are present. Disc weeds under and level any irregularities in the grade, especially when planning furrow irrigation. If perennials, such as johnsongrass or bermudagrass, grow on the site, they are easier and less expensive to control before you plant the trees. During early fall, when the perennials are still flowering, treat with glyphosate; repeat the treatment in the spring to kill regrowth, and disc about 10 days later to expose the root system for drying. Before planting, usually in the spring, a preemergence herbicide such as trifluralin can be incorporated over the entire site or into 4- to 6-foot strips where the trees will be planted. Napropamide can be applied after planting. Young citrus trees are less susceptible to trifluralin and napropamide than to other preemergence herbicides. Such a preemergence application usually controls germinating seedlings throughout the summer.

Once trees are planted, disturb the soil as little as possible if you plan nontill management. For furrow irrigation, establish one or two narrow furrows along the planted trees. Spot or broadcast treatment of perennials with contact herbicides may be needed, depending on the extent of preplant control of the perennials. In late fall, just before the first rains, apply a preemergence herbicide to control the winter weeds you expect from your sping survey. Regular preemergence and postemergence treatments during the establishment years remove much of the competition by weeds and facilitate irrigation and other cultural operations.

Established Orchards. In established orchards, weed management has to be adjusted to the irrigation method used. In orchards irrigated by furrows, weeds are a particular problem in furrow bottoms and at furrow ends where high moisture and extensive leaching of herbicides allow weeds to grow. Where herbicides are applied with cluster nozzles, one side of the furrows is in the "shadow" of the spray stream and remains untreated. Under a low-volume irrigation scheme, the permanently wet zone around the emitters or sprinkler heads favors weed growth and promotes the breakdown of soil-applied herbicides. In each case, postemergence treatment of escaping weeds is often needed. When furrow or drip irrigation is used, herbicides must be applied before a rain to carry the material into the soil. In orchards irrigated by regular sprinklers, low-volume sprinklers, or flooding, preemergence herbicides can be applied anytime before an irrigation.

Annual herbicide treatments—usually one fall or early winter application or a split application between fall and spring—are generally needed in a nontilled orchard, but the rates usually can be lowered over the years. During the summer, some cleanup of summer annuals or perennials may be necessary, particularly in moist areas. The objective of weed control, however, is not to eliminate weeds but to keep their growth at a minimum and to avoid the buildup of resistant species. This can be achieved by carefully selecting herbicides or herbicide combinations, splitting applications, and rotating herbicide materials.

Weed Species Common in Citrus Orchards

The spectrum of weed species found in citrus orchards varies greatly, depending on climate, soil, and cropping history. Some of the most common weed species in citrus are illustrated below; for other species, consult the University of California *Growers Weed Identification Handbook* #4030, Agricultural Sciences Publications, or your local farm advisor.

Bermudagrass
Cynodon dactylon

Bermudagrass, commonly used as turf grass, is troublesome in orchards mainly in southern California. It is a perennial that reproduces from rhizomes, stolons and seeds. The root system tolerates salt and drought but is killed when exposed to the sun. A preemergence treatment for the seedlings followed by a postemergence application of glyphosate provides good control. Glyphosate is most effective when the bermudagrass is blooming or has just formed seed heads.

A.

B.

A. The mature plant forms dense mats with spreading and branching stolons. Erect stems arise from the axils of the short leaves on the stolons; the stems are 10 to 45 cm (4 to 18 inches) tall and have longer leaves than the stolons. The flower head consists of three to seven slender spikes radiating from one point at the tip of the stem. The flower head of large crabgrass (*Digitaria sanguinalis*) resembles that of bermudagrass, but the spikes usually arise slightly apart on the stem. As an annual plant, crabgrass does not have stolons or an extensive rootstock.

B. The collar region of bermudagrass has a fringe of short, white hairs and often has tufts of long hairs on the margins; auricles are absent.

Dallisgrass
Paspalum dilatatum

Dallisgrass is occasionally troublesome in orchards under limited tillage and frequently in areas with standing water. It is a bunchgrass and does not produce stolons. The rhizomes of dallisgrass are fairly short and have short internodes, forming concentric rings. To control dallisgrass, drain wet areas and spot treat established plants with glyphosate or MSMA; seedlings are controlled by most preemergence herbicides.

C. The mature plant forms coarse, loose bunches, 30 to 120 cm (1 to 4 feet) high. The flower head consists of 3 to 6 spikes, which are spaced apart and are often drooping.

D. The leaf sheath of dallisgrass is somewhat flattened; at the base, it is very hairy, often tinged red and usually inflated. The seedling and young plant often grow prostrate.

E. The collar region has a firm, membranous ligule and a few spreading hairs at the margins; auricles are absent.

C.

D.

E.

Field bindweed
Convolvulus arvensis

Field bindweed or morningglory is a perennial that reproduces from an extensive rootstock and abundantly produced seeds. The seeds may remain dormant for many years. Field bindweed is best controlled before planting the orchard. The preemergence herbicide trifluralin controls germinating seedlings. Glyphosate can be used as a postemergence treatment. Disc the site first, then irrigate

F.

G.

to stimulate vigorous regrowth of the bindweed. Treat the regrowth with glyphosate when a few flowers are present and disc the site again after 3 or 4 weeks to destroy the seedlings.

F. The mature plant grows prostrate with slender stems, 1 m (3 feet) or longer, or upward while entwining other plants. The flowers are funnel shaped, white to light red and are borne singly on slender stalks in leaf axils. They open only on sunny mornings.

G. The seed leaves are broad and notched at the tip. The leaf stalks are grooved above. The first true leaves are heart shaped and deeply lobed at the base; later leaves are generally shaped like an arrowhead. Young plants beyond the five-leaf stage begin to form their extensive root system and are very difficult to control.

Nutsedges
Cyperus spp.

Nutsedges are perennial weeds that infest many annual and perennial crops. Of the two species found in California citrus orchards, yellow nutsedge (*C. esculentus*) is widespread in the central valleys, whereas purple nutsedge (*C. rotundus*) is more common to parts of southern California. The two species differ in their appearance but are controlled in the same way.

Nutsedges can reproduce from seed but apparently spread mainly through abundantly produced tubers or nuts. Plants produce tubers on rhizomes as deep as 20 cm (8 inches); the tubers remain viable for several years, even in dry soil. When conditions are favorable, the tubers grow a new system of roots, rhizomes, and aboveground shoots. Long-term control of nutsedge requires preventing the plants form growing beyond the 5- to 6-leaf stage, that is, destroying them before they start producing new tubers. Nutsedges thrive in sunny, moist areas; in the shade, their growth is reduced.

H. Nutsedges grow to a height of 30 to 60 cm (1 to 2 feet). Flowering stalks are triangular in cross section and usually are no longer than the basal leaves in yellow nutsedge (shown here); in purple nutsedge, the flowering stalks are longer than the basal leaves. Young nutsedge plants are grasslike, but the leaves are thicker and stiffer than most grasses. The leaves are triangular in cross section and are arranged in a spiral at the base.

I. The tubers of yellow nutsedge are round, light brown with a smooth surface and have a pleasant, nutlike flavor. They form singly on rhizomes. Tubers of purple nutsedge are elongated, reddish, coarse, and have a bitter flavor. They form in chains on rhizomes.

H.

I.

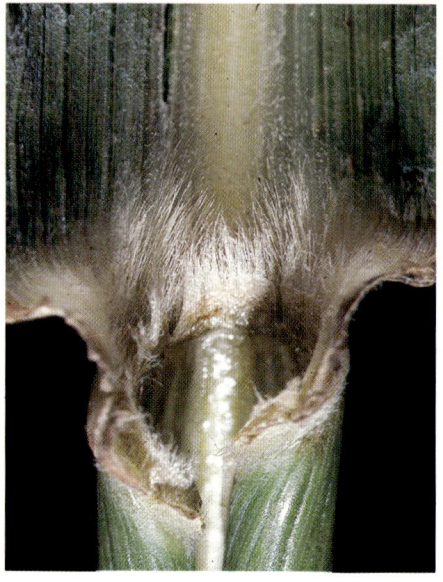

J.

K.

L.

Johnsongrass
Sorghum halepense

Johnsongrass is a common perennial grass in orchards, particularly in the San Joaquin Valley. It thrives in annual crops and along ditchbanks, where it is often spread by irrigation water. Like dallisgrass, it does not form stolons but reproduces from rhizomes and seeds. The seeds may be dormant for many years. Most preemergence herbicides control the seedlings. A spot treatment of established plants with MSMA or glyphosate may occasionally be needed.

J. The mature plant grows in stout, leafy patches, which may be up to 2 m (6 to 7 feet) tall. The leaves have a prominent whitish midvein, which snaps readily when folded over. The leaf sheath is somewhat flattened and reddish at the soil line but is not hairy, as in dallisgrass. The young plant may resemble barnyardgrass, but barnyardgrass does not have a ligule.

K. On mature stems, the ligule has a toothed margin and a fringe of fine hairs; young stems have a thin, papery ligule with torn margins.

L. The rhizomes are thick, fleshy, and white to purplish; they are longer and have longer internodes than those of dallisgrass.

Bearded Sprangletop
Leptochloa fascicularis

Bearded sprangletop, also known as sprangletop, is a summer annual grass that is commonly found throughout the central valley and is usually associated with alkaline soils. It thrives in wet places, such as the ends of furrows and around stand pipes. Provide good drainage in those areas to discourage weed growth.

M. The mature plant is coarse, upright growing and forms large tufts 30 cm to 1 m (12 to 40 inches) tall. The flower heads are long, erect, well branched, and straw colored at maturity.

N. The collar region has a thin, long, and delicate ligule and no auricles.

M.

Barnyardgrass
Echinochloa crus-galli

Several varieties of barnyardgrass or watergrass grow in California; they differ mainly in the appearance of the flower head. Diuron or a diuron/bromacil mixture control barnyardgrass.

O. The mature plant grows upright, 15 cm to 2 m (6 inches to 6 feet) tall, often forming dense tufts. The leaf sheath is flattened and reddish tinged at the soil line. The lower spikes of the flower head are spaced apart; the top spikes are crowded together. The flower groups are barbed with short, stiff hairs or long bristles, depending on the variety.

P. Barnyardgrass is the only major summer weedy grass without a ligule.

N.

O.

P.

Spotted Spurge
Euphorbia maculata

Q.

R.

Spotted spurge belongs to the spurge family, whose members have a milky, sticky sap and small, inconspicuous flowers. Spotted spurge or prostate spotted spurge is a summer annual weed which is suitable as a ground cover because it grows low to the ground and is much less competitive with trees than other annuals and perennials. Where spurge occurs, it is easy to establish as a ground cover; it is resistant to commonly used herbicides, such as simazine and diuron, and thrives in the absence of other weed competitions.

Q. The mature plant is prostrate or low growing. Leaves grow opposite on short stalks, are finely toothed on the margins, and are covered with soft hairs. The plant got its name from the dark red spot often found in the middle of its leaves.

R. The seed leaves are oval, about one- and one-half to twice as long as they are wide, with a rounded tip and smooth margins. They are bluish green, have a powdery bloom on the upper surface and a reddish tinge underneath. The first true leaves are more rounded at the tip, tapered toward the leaf stalk, and show the reddish spot on the leaf middle.

Turkey Mullein
Eremocarpus setigerus

S.

T.

Turkey mullein, is a member of the spurge family and a native summer annual which is widely distributed throughout California. It grows in open, dry areas and has become prevalent in citrus orchards because it is resistant to commonly used herbicides, such as diuron and simazine.

S. The mature turkey mullein plant is low growing, up to 30 cm (1 foot) high, and forms dense, rounded mats 30 to 60 cm (1 to 2 feet) across. Stems and leaves of the evenly branched plant are densely covered with bristly, star-shaped hairs. The leaves are thick, oval to nearly round, three-nerved, and have a strong, sweet odor. The flowers are small and inconspicuous; the male flowers have six or seven stamens and arise in clusters at the tip of branches; the female flowers, with a single, densely haired ovary and style, are borne in lower leaf axils.

T. The seed leaves and true leaves are oval and covered with star-shaped hairs, which give the plant a gray green, mealy appearance.

References

General
Citrus Growing in California. Publication 4011.*

Citrus Growing in the Sacramento Valley. Leaflet 2443.*

The Citrus Industry, Vol. II (Anatomy, Physiology, Genetics, and Reproduction) W. Reuther, L. D. Batchelor, and H. J. Webber (eds.). University of California, Division of Agricultural Sciences, 1968, 398 pp.*

The Citrus Industry, Vol. III (Production Technology). W. Reuther (ed.). University of California, Division of Agricultural Sciences, 1973, 528 pp.*

The Citrus Industry, Vol. IV (Crop Protection). W. Reuther, E. C. Calavan, G. E. Carman (eds.). University of California, Division of Agricultural Sciences, 1978, 362 pp.*

Horticulture
Basic Irrigation Scheduling. Leaflet 21199.*

Drip Irrigation. Leaflet 2740.*

Preharvest Use of 2,4-D on Citrus. Leaflet 2447.*

Protecting Citrus from Cold Losses. Leaflet 2372.*

Pruning Citrus Trees. Leaflet 2449.*

Questions and Answers about Tensiometers. Leaflet 2264.*

Pesticides
Calibration of Herbicide Sprayers. Leaflet 2710.*

How Much Chemical Do You Put in the Tank? Leaflet 2718.*

Pesticide Application and Safety Training Manual. Publication 4070.*

Pesticide Toxicities. Leaflet 21062.*

Reducing Pesticide Hazards to Honey Bees with Integrated Management Strategies. Leaflet 2883.*

Safe Handling of Agricultural Pesticides. Leaflet 2768.*

Treatment Guide for California Citrus Crops, 1984–1986. Leaflet 2903.*

Vertebrates
Guide to Vertebrate Control Materials Registered in California. Leaflet 21226.*

Pocket Gopher Control with Mechanical Bait Applicator. Leaflet 2699.*

Vertebrate Pest Control Handbook. California Department of Food and Agriculture, 1975.

Insects and Mites
A Portable Apparatus for Detection of Virus-Diseased Citrus Red Mites in the Field. J. Economic Entomology, June 1972, pp. 890–891.

Diseases
Color Handbook of Citrus Diseases. L. J. Klotz. University of California, Division of Agricultural Sciences, 1973, 121 pp.*

Nematodes
General Recommendations for Nematode Sampling. Leaflet 21234.*

Weeds
Growers' Weed Identification Handbook. Publication 4030.*

Weed Management Guide for Citrus. Leaflet 2979.*

Weeds of California. W. W. Robbins, M. K. Bellue, and W. S. Ball. State of California Documents and Publications, P.O. Box 1015, North Highland, CA 95660.

---

*These publications are available from the University of California, Agricultural Sciences Publications; see p. 2 for address. A free catalog lists many other titles on agricultural topics.

Glossary

albedo. white, spongy inner part of citrus fruit rind.

chlorosis. a diseased condition in which normally green plant tissue is yellowed or bleached.

cocoon. a sheath, usually mostly of silk, formed by an insect larva as a chamber for pupation.

cortex. in stems or roots, the tissue between the epidermis and conducting tissue.

crawler. the active first instar of a scale insect.

cultivar. a cultivated plant variety.

diapause. a period of physiologically controlled dormancy in insects.

epidermis. the outermost layer of cells on the bodies of animals or on plant surfaces.

evapotranspiration. the loss of water from the soil by the combination of surface evaporation and transpiration by plants.

flavedo. outer part of the rind of citrus fruit, bearing oil glands and pigments.

girdle. to kill or damage a ring of tissue around a stem or root; such damage interrupts the transport of water and nutrients.

honeydew. an excretion from insects, such as aphids, mealybugs, and soft scales, consisting of modified plant sap.

indexing. testing a plant for infection, usually by grafting tissue from it onto an indicator plant.

instar. the stage of an insect between successive molts.

larva. the immature form of an insect, such as a caterpillar or maggot, that hatches from an egg and passes through a pupal stage before becoming an adult; for nematodes, any life stage between the embryo and adult.

lesion. a well-defined area of diseased tissue, such as a canker or leaf spot.

meconium. fecal pellet excreted by a larva before pupation.

mutation. the abrupt appearance of a new, heritable characteristic as the result of a change in the genetic material of one individual.

mycelium. the vegetative body of a fungus, consisting of a mass of slender filaments or hyphae.

mycoplasma. a member of the genus *Mycoplasma*. Mycoplasmas, unlike viruses, can reproduce in the absence of a host and are the smallest free-living organisms; they have a unit membrane but no cell wall as do bacteria.

necrosis. death, usually of a well-defined part of a plant such as a leaf or the areas inside a canker.

nymph. the immature stage of insects such as grasshoppers and aphids that hatch from eggs and gradually acquire adult form through a series of molts without passing through a pupal stage.

parasite. an organism that lives in or on the body of a larger living organism (its host); in this book the term also refers to insect parasitoids, which spend their immature stages on or within the body of a single host, killing it just before pupation.

pathogen. any disease-producing organism.

pheromone. a substance secreted by an organism to affect the behavior or development of other members of the same species; sex pheromones that attract the opposite sex for mating are used in monitoring certain insects.

phloem. food-conducting tissue, consisting mostly of sieve tubes; in aboveground parts of the tree it forms a delicate layer between the bark and water-conducting vessels (xylem or wood); in roots, it is organized with the xylem in the central stele.

phytotoxic. causing injury to plants.

pulp. juice-containing vesicles underneath the rind of citrus fruit.

pupa. a nonfeeding, inactive stage in which the tissues of an insect larva are reorganized into those of an adult.

resistant. able to withstand conditions harmful to other strains of the same species.

rootstock. the lower portion of a graft that grows into the root system.

scion. the upper portion of a graft that grows into the trunk or branch.

sepal. one of the outermost flower structures which usually enclose the other flower parts in the bud (Figure 2).

sieve tubes. see phloem.

sooty mold. a sooty coating on foliage or fruit, formed by the dark mycelia of fungi that live in the honeydew secreted by certain insects.

spore. a reproductive body produced by certain fungi and other organisms, capable of growing into a new individual under certain conditions.

stele. the central cylinder inside the cortex of many roots and stems.

tolerant. able to withstand the effects of a condition without suffering serious injury or death.

viroid. a portion of infectious nucleic acid, without the protein coat of a virus.

virus. a small parasitic organism, consisting only of nucleic acid and a protein coat, that can reproduce only within the living cells of a host.